Modern Permanent Magnets - Fundamentals and Applications

Edited by Dipti Ranjan Sahu

Published in London, United Kingdom

Modern Permanent Magnets - Fundamentals and Applications
http://dx.doi.org/10.5772/intechopen.107553
Edited by Dipti Ranjan Sahu

Contributors
Alfredo Bermúdez, Belqees Hassan, Daniel Benjamín Berdichevsky, Dipti Ranjan Sahu, Dolores Gómez, Pablo Venegas, Tingting Yin

First published in London, United Kingdom, 2024 by IntechOpen
IntechOpen is the global imprint of INTECHOPEN LIMITED, registered in England and Wales, registration number: 11086078, 5 Princes Gate Court, London, SW7 2QJ, United Kingdom

British Library Cataloguing-in-Publication Data
A catalogue record for this book is available from the British Library

Additional hard and PDF copies can be obtained from orders@intechopen.com

Modern Permanent Magnets - Fundamentals and Applications
Edited by Dipti Ranjan Sahu
p. cm.
Print ISBN 978-1-83769-044-2
Online ISBN 978-1-83769-045-9
eBook (PDF) ISBN 978-1-83769-046-6

Meet the editor

Prof. Dipti Ranjan Sahu is a Professor of Physics in the Department of Biology, Chemistry and Physics, Namibia University of Science and Technology (NUST). He received his Ph.D. in Physics from the Institute of Materials Science, Utkal University, India. After completing his Ph.D., Dr. Sahu worked as a postdoctoral researcher and visiting scientist/faculty member at several institutions, such as National Taiwan University, National Cheng Kung University, Taiwan, and the University of Witwatersrand, South Africa. His research focus is on multifunctional materials including nanomaterials, ceramics, composites, spintronics, and magnetic materials and the application of functional materials in devices. He has published more than 115 peer-reviewed articles, 6 books, 6 book chapters, and more than 110 research articles in conference proceedings and meetings.

Contents

Preface

Permanent magnets are powerful magnets that maintain their magnetic properties for a long time. Modern permanent magnets are important for academic research, defense applications, energy technologies, and more. There are several types of permanent magnets, each manufactured differently from different materials with different properties. The method of manufacturing and fabrication of permanent magnets is important for the design and development of new devices and their assembly into functional systems for different applications in different fields. The material used for making a permanent magnet also possesses specific properties such as the high value of coercivity and retentivity; the area of the magnet hysteresis curve should also be large. Permanent magnets can be bonded magnets or polymer-free magnets, with rare-earth materials or without them.

This book presents an update and overview of the principles and recent developments in various families of permanent magnets. The chapters discuss permanent magnets and their manufacturing processes. They also present the latest research developments in different types of modern permanent magnets, scientific advances in manufacturing techniques, and the applications of magnets, providing insight into the development of future devices.

This book contains six chapters. Chapter 1 introduces the topic of permanent magnets. Chapter 2 discusses light–matter interaction in 2D magnets. Chapter 3 explains the environmental impact of permanent magnets. Chapters 4 and 5 deal with the hysteresis model and magnetized plasma. Finally, Chapter 6 presents recent advances in nanocrystalline permanent magnets of different materials.

This book is a useful resource for students, researchers, academicians, and professionals. It presents a broad range of ideas and proposes possible solutions for the development of effective permanent magnets.

Dipti Ranjan Sahu
Faculty of Health,
Department of Biology,
Chemistry and Physics,
School of Natural and Applied Sciences,
Natural Resources and Applied Sciences,
Namibia University of Science and Technology,
Windhoek, Namibia

Chapter 1

Introductory Chapter: Modern Permanent Magnets – Basics and Applications

Dipti Ranjan Sahu

1. Introduction

Permanent magnets are highly magnetized functional hard materials, which do not lose magnetism over time due to the generation of magnetic field by the internal structure of the material itself [1–3]. These modern permanent magnets are made from a "cocktail" of minerals which can include iron, neodymium, samarium, cobalt and nickle. Normally, the functionality of the permanent magnets depends on the intrinsic properties of the base compound. It is commonly known that permanent magnets are source of magnetic field. The composition, extrinsic properties such as microstructure and processing method of materials are define the magnetic properties of the permanent magnets [4–6].

2. Basics of permanent magnets

The magnetic flux with no energy input defines the uniqueness of permanent magnets. The performance of permanent magnets is estimated based on the magnetization (M) and maximum energy product and the magnetic parameter such as $(BH)_{max}$, spontaneous magnetization and coercive forces [7].

In addition, to above indicator, some basic requirements of permanent magnets are mentioned as follows:

- Anisotropy

Shape anisotropy, magneto-crystalline anisotropy and stress anisotropy contribute towards the anisotropy behavior of permanent magnets [8].

- Coercive force/coercivity

High coercivity is the requirement for the development of permanent magnets, which can be achieved by controlling the microstructure of the major constituent phase of the magnetic compound [8, 9].

- Magnetic domains and domain walls

Magnetic domain observations are critical for permanent magnet. The thickness of domain wall and domain wall structure inform the change in magnetic moment and the formation of permanent magnets [10].

- Magnetic hysteresis

This is an important tool for the quantitative analysis of permanent magnet performance. Hard magnetic materials always show the greatest hysteresis [11].

- Recoil permeability:

The recoil permeability describes the steepness of the demagnetization curve in the $B(H)$ explanation [12].

- Stability

Excellent magnetic performance of permanent magnets materials: Materials must possess large exchange interactions, high spontaneous magnetizations and large magnetic anisotropy [13].

Overall, for a permanent magnet, the material must attribute to high saturation magnetisation (M_s), Curie temperature (T_c) and magnetocrystalline anisotropy (MAE).

3. Types of modern permanent magnets and materials

Modern permanent magnets are manufactured from different types of materials and systematic variations of compounds using different process. Based on composition variations in compounds, sub-lattice substitutions, crystallization of amorphous deposits and structure, permanent magnets are classified.

Generally, five different types of permanent magnets are known and produced for different purposes [14–16]. They are as follows:

i. Aluminium-nickel-cobalt (Alnico)

ii. Ferrite

These are M-type Sr- and Ba-hexaferrites

iii. Samarium-cobalt (SmCo)

Further in the Sm-Co family, again two types of permanent magnet such as $SmCo_5$ and $Sm_2(Co,Fe,Cu,Zr)_{17}$ type are based on the variation of compositions and substitution of different elements.

iv. Neodymium-iron-boron (NdFeB) magnets

Similarly, NdFeB magnets are classified into sintered NdFeB (main phase is $Nd_2Fe_{14}B$), hot-deformed NdFeB magnets, bonded NdFeB produced using melt spun powder and bonded NdFeB magnets made from hydrogenation-disproportionation-desorption-recombination powder.

v. SmFeN

SmFeN magnet, also known as $Sm_2Fe_{17}N_3$, is a potential option for next generation of permanent magnetic material.

4. Key challenges of modern permanent magnets

Design and development of permanent magnets have challenges for the intrinsic and extrinsic magnetic effects due to mechanical, thermal and chemical parameters [1]. The challenges like metallurgical processing, microstructural changes and phase stability understanding are important for the development of good permanent magnets [17, 18]. The basic properties of the magnets such as remanent magnetization and coercivity are dependent on microstructure [7]. The parameters like room temperature $(BH)_{max}$, M_s and H_c, J diminish rapidly with increasing temperature. The fine grain size also has effect on the coercivity at high temperature [19]. Further, the packaging of magnet in operating environment should be thermodynamically stable, so that magnets will be physically stable (resistant to corrosion, decomposition or oxidation). Again, challenges differ from one another on the selection of material system, which show simultaneous high magnetization and high coercivity at any given time and environment.

5. Application of modern permanent magnets

Application of permanent magnets are many folds, starting from mechanical instruments such motors, generators, sensors, loudspeakers, bearings and clutches, automobiles to missiles, electrical to mechanical energy conversion, traveling wave tubes, transmission and distribution of electric power, microwave communications, and data storage [20]. Permanent magnets are a very crucial component of numerous industries and markets such as transportation, industrial, residential/commercial, consumer electronics, defence, IT/PC, wind energy and medical [21]. Permanent magnets are also important for military application, which add functionality to jet fighter engines and electronic countermeasure, missile and satellite communication systems. Many energy-efficient appliances, electric vehicles and direct-drive wind turbines also use permanent magnets [1, 22, 23].

Based on the application of permanent magnets, different materials are used for the manufacture of suitable permanent magnets, which can work at high temperature and harsh environment keeping the strength of the magnet intact.

6. Future of modern permanent magnets

The permanent magnet's role is to provide a magnet field in a substance or vacuum for specific applications. Development of material and technology improved the appropriate need, functionality and flexibility of permanent magnets. Research progress in nanomaterial [24–26] and nanotechnology area meet the future demand of permanent magnet and miniaturization of devices. Further new development in rare earth permanent magnets opens the field of future permanent magnet markets, which plays an important role in a carbon neutral society. Understanding of cost-effective practical manufacturing process and material incorporation for effective production of permanent magnets contribute to a brighter future for permanent magnet. Following consideration can be taken for the development of new modern permanent magnets, which can be commercialized for future markets [27–29].

A. Improvement of magnet properties

a. Enhance the existing magnetic material types: manipulation of crystal structure, increased the coercivity and H.

b. Materials modification for specific requirement: addition of new materials or change of composition of existing materials.

c. Improve the magnet manufacturing technology.

d. Develop user friendly magnets.

B. New magnetic materials for modern magnets

Search for new magnetic materials having high Curie point, easy-axis-type crystal anisotropy and high saturation magnetization. Nanomaterials combine with the used permanent materials may be an option for new ferromagnetic materials.

C. Manipulation of permanent magnet properties and calculation of theoretical limit

Look for high H magnetic materials, then theoretical calculation can be done ignoring coercive force and anisotropy. It is reported that rare earth addition additions in transition-metal alloys can improve the magnetic properties. Similar avenue can be search for the development of new permanent magnets.

D. Sensitivity of magnet performance

Processing conditions affect magnet manufacturing costs. Cost structures determine the commercialization process. Therefore, proper care should be taken staring from synthesis pathways to application. New tools, technique, experimental method and computational work can provide detail description about the performance of the magnets.

7. Summary

Field of permanent magnets is diverse. Proper research and attention create a large demand in this area for the development of variety of materials as per the application requirements. The existing permanent magnets with superior performance and development of new permanent magnets are the future of the new devices and industry.

Author details

Dipti Ranjan Sahu
Department of Biology, Chemistry and Physics, School of Natural and Applied Sciences, Namibia University of Science and Technology, Windhoek, Namibia

*Address all correspondence to: dsahu@nust.na

References

[1] McCallum RW, Lewis LH, Skomski R, Kramer MJ, Anderson IE. Practical aspects of modern and future permanent magnets. Annual Review of Materials Research. 2014;**44**:451-477

[2] Yazid MM, Olsen SH, Atkinson G. MFM study of a sintered Nd–Fe–B magnet: Analyzing domain structure and measuring defect size in 3-D view. IEEE Transactions on Magnetics. 2016;**52**(6):1-10

[3] Narasimhan K. Fundamental properties of permanent magnets. In: Croat JJ, Ormerod J, editors. Woodhead Publishing Series in Electronic and Optical Materials. 1st ed. USA; 2022. pp. 31-63. ISBN-13:978-0323886581

[4] Chen Z, Wang F, Yan S, et al. Microstructure and magnetic properties of M-type $Sr_{0.61-x}La_{0.39}$ $Ca_xFe_{11.7}Co_{0.3}O_{19}$ hexaferrite prepared by microwave calcination. Materials Science and Engineering B. 2014;**182**:69-73

[5] Coey JMD. Permanent magnets: Plugging the gap. Scripta Materialia. 2012;**67**:524-529

[6] Folks L, Woodward RC. The use of MFM for investigating domain structures in modern permanent magnet materials. Journal of Magnetism and Magnetic Materials. 1998;**190**:28-41

[7] Costa EC, Lima AO, dos Santos MNL, Souza RM, Wendhausen PAP, Xavier FA. On the mechanical behavior of the sintered Nd-Fe-B permanent magnet during diamond scratching. Journal of Manufacturing Processes. 2023;**94**:79-93

[8] Lewis LH, Jim'enez-Villacorta F. Perspectives on permanent magnetic materials for energy conversion and power generation. Metallurgical and Materials Transactions A: Physical Metallurgy and Materials Science. 2013;**44**:2-20

[9] Sepehri-amin H, Hirosawa S, Hono K. Advances in Nd-Fe-B based permanent magnets. In: Handbook of Magnetic Materials. 1st ed. Vol. 27. Elsevier B.V.; 2018. pp. 269-372

[10] Makoto T, Suzuki M, Kobayashi S, Kotani Y, Nakamura T, et al. Real picture of magnetic domain dynamics along the magnetic hysteresis curve inside an advanced permanent magnet. NPG Asia Materials. 2022;**14**:70

[11] Coey JMD. Hard magnetic materials: A perspective. IEEE Transactions on Magnetics. 2011;**47**:4671-4681

[12] Müller KH, Sawatzki S, Gauß R, Gutfleisch O. Permanent magnet materials and applications. In: Coey JMD, Parkin SS, editors. Handbook of Magnetism and Magnetic Materials. Cham: Springer; 2021. pp. 1369-1433

[13] Bi X, Qiu C, Qin F, Huang J, Yuan H. Exchange interactions and emergent magnetic phenomena in layered *3d*-orbital ferromagnets. Advanced Physics Research. 2023;**2**:22001061 (1-21)

[14] Croata JJ, Ormerod J. The history of permanent magnets. In: Croat JJ, Ormerod J, editors. Woodhead Publishing Series in Electronic and Optical Materials. 1st ed. USA; 2022. pp. 3-29. ISBN-13:978-0323886581

[15] Hedlund D. Digital Comprehensive Summaries of Uppsala Dissertations from the Faculty of Science and Technology 2160 Uppsala University, SE-751 03 Uppsala, Sweden. ISSN 1651-6214

[16] Delette G. $Nd_2Fe_{14}B$ permanent magnets substituted with non-critical light rare earth elements (Ce, La): A review. Journal of Magnetism and Magnetic Materials. 2023;**577**:170768 (1-16)

[17] Richman RH, McNaughton WP. Permanent-magnet materials: Research directions and opportunities. Journal of Electronic Materials. 1997;**26**(5):415-422

[18] Hono K, Sepehri-Amin H. Strategy for high-coercivity Nd–Fe–B magnets. Scripta Materialia. 2012;**67**:530-535

[19] Cui J, Ormerod J, Parker D, Ott R, et al. Manufacturing processes for permanent magnets: Part I–sintering and casting. Advances in Processing, Manufacturing, and Applications of Magnetic Materials. JOM. 2022;**74**(4):1279-1295

[20] Gupta A, Roy PK. Effect of Zn^{2+} ion substitution in Al^{3+}-substituted rare-earth free Sr-hexaferrite for different permanent magnet applications. Inorganic Chemistry Communications. 2023;**155**:111114

[21] Ormerod J. Permanent magnet markets and applications, modern permanent magnets. In: Woodhead Publishing Series in Electronic and Optical Materials. 2022. pp. 403-434

[22] Sugimoto S. Current status and recent topics of rare-earth permanent magnets. Journal of Physics D: Applied Physics. 2011;**44**:064001

[23] Yamazaki K, Fukushima Y. Effect of eddy-current loss reduction by magnet segmentation in synchronous motors with concentrated windings. IEEE Transactions on Industry Applications. 2011;**47**:779-788

[24] Hirosawa S et al. Perspectives for high-performance permanent magnets: Applications, coercivity, and new materials. Advances in Natural Sciences: Nanoscience and Nanotechnology. 2017;**8**:013002 (1-12)

[25] Hao S-Z, Zhou D, Pang L-X, et al. Ultra-low temperature co-fired ceramics with adjustable microwave dielectric properties in the Na_2O-Bi_2O_3–MoO_3 ternary system: A comprehensive study. Journal of Materials Chemistry C. 2022;**10**:2008-2016

[26] Li D, Zhou D, Wang D, Zhao W, et al. Lead-free relaxor ferroelectric ceramics with ultrahigh energy storage densities via polymorphic polar nanoregions design. Small. 2023;**19**:2206958

[27] Coey MD, Iriyama T. Bonded Sm-Fe-N permanent magnets, modern permanent magnets. In: Woodhead Publishing Series in Electronic and Optical Materials. 2022. pp. 305-342

[28] Yoshida Y, Yoshikawa N. Hot formed NdFeB magnets, modern permanent magnets, modern permanent magnets. In: Woodhead Publishing Series in Electronic and Optical Materials. 2022. pp. 251-304

[29] Hioki K. High performance hot-deformed Nd-Fe-B magnets (Review). Science and Technology of Advanced Materials. 2021;**22**(1):72-84

Chapter 2

Novel Light-Matter Interactions in 2D Magnets

Tingting Yin

Abstract

Since the discovery of intrinsic long-range magnetic order in two-dimensional (2D) layered magnets, e.g., $Cr_2Gr_2Te_6$ and CrI_3 in 2017, it has attracted intensive studies of new physical phenomena in these systems down to a few atomic layers, especially, their magnetism ground states at finite temperatures. Recently, the light-matter interactions in 2D magnets, including light absorption, emission, scattering, et al., have gradually drawn researchers' attention and are current active research directions. The mechanism of light-matter interactions in 2D magnets challenges the knowledge of materials physics, which drives the rapid development of materials synthesis and device applications. In this chapter, an overview of crystal structures, magnetic properties, and electronic band structures is presented. More importantly, the current status of light-matter interactions in 2D magnets will be discussed, which provides a solid basis for understanding novel physical phenomena in 2D magnets and proves the importance of tuning the magnetic, electronic, and vibrational degrees of freedom for designing novel 2D magnet-based device applications.

Keywords: 2D magnets, light-matter interaction, magnetism, light emission, light scattering

1. Introduction

Materials with reduced dimensionality have become particularly important, owing to their unexpected physical and chemical properties. Since the discovery of graphene, fabricated via the Scotch tape exfoliation method from graphite in 2014 [1, 2], scientists have made intensive efforts to search two-dimensional (2D) van der Waals (vdW) materials. This family includes metals, e.g., $NbSe_2$, semiconductors, e.g., transition-metal dichalcogenides (TMDs), insulators, e.g., hexagonal boron nitride (hBN), and other elemental 2D vdW materials [3–5]. In particular, exfoliated ultrathin 2D vdW materials down to atomic thickness (one or several unit layers) display unique and fascinating electronic and optical properties. For example, optical excitation on the monolayer TMDs will generate a strongly bound electron-hole pair called an exciton, instead of fee electrons and holes as in traditional bulk semiconductors [6]. Moreover, their extraordinary mechanical property makes these atomic thin materials with breakthrough of integration technologies accelerate the on-chip device applications [7].

Magnetism in two dimensions has been long pursued by scientists. 2D magnets are at the heart of numerous theoretical [8, 9], experimental [10, 11] and technological advances [11], which are considered as one of the most promising solutions for next-generation spintronic devices in a revolution in semiconductor technology [12–14]. However, 2D vdW materials with intrinsic magnetism had been missing until quite recently. Based on the theoretical point of view, the transition of magnetism from bulk to 2D, the fluctuations of any kind will increase and can easily destroy long-range magnetic order at any finite temperature [15]. In addition, the superexchange will also be modified with the dimension reduction of the system due to the breaking of the inversion symmetry or the modification of the electronic/lattice structure [16]. Thus, the magnetic properties of 2D vdW materials are far less experimentally investigated for a long time.

With the first discovery of intrinsic magnetic order in the monolayer/few-layer limit in 2D ferromagnets, $Cr_2Ge_2Te_6$ and CrI_3 in 2017 [17, 18], this trend may be poised to change. Subsequently, various 2D vdW magnets have been rapidly discovered and studied [19]. In contrast to the conventional magnetic materials, the magnetic order of 2D magnets can persist down to the monolayer limit, thus possessing a vast reservoir of properties. For example, (i) 2D magnets have strong quantum confinement and mechanical flexibility [20]; (ii) 2D magnets possess good sensitivity to external fluctuations (doping, strain, electric field, et al.) [21]; (iii) 2D magnets can form various artificial heterostructures by a simple stacking process [22]; (iv) 2D magnets show thickness-dependent tuning properties [23]. Thus, 2D magnets have been demonstrated promising building blocks for next-generation information devices. Currently, intense experimental and theoretical investigations are almost devoted to explore the intrinsic magnetism, increase Curie temperatures, and design/fabricate 2D magnet-based devices with advanced functionalities. However, the investigation of light-matter interactions in these 2D magnets is still at the primary stage, there have already been reported unique optical properties in some ferromagnetic (FM) or antiferromagnetic (AFM) ones, exhibiting quite different light emission/scattering as compared with conventional nonmagnetic 2D materials. The mechanism of light-matter interactions in 2D magnets challenges the knowledge of materials physics, which drives the rapid development of materials synthesis and device applications.

In this chapter, the special topic of light-matter interactions in atomically thin 2D magnets is mainly discussed. We begin with a brief review of the recent remarkable progress of achieved 2D vdW magnets, mainly on CrX_3 (X = Cl, Br, I), MPS_3 (M = Fe, Ni, Mn), CrSBr and $CrPS_4$, with special emphasis on the magnetic properties and electronic band structures by changing the number of layers. Then, we mainly discuss light absorption, emission and scattering behaviors in 2D magnets. Additionally, the mechanism of light-matter interactions in 2D magnets determined by complex interactions between magnetic, electronic and vibrational degrees of freedom will be elaborated in detail, which plays a key role in the state-of-the-art of opto-electronic device applications.

2. Recently developed 2D magnets

In recent years, a great deal of new 2D magnets have been experimentally discovered and theoretically predicted. Here, we classified these emerging 2D intrinsic magnets into binary transition metal halogenides: CrX_3 (X = Cl, Br, I) and ternary transition metal compounds, MPS_3 (M = Fe, Ni, Mn), CrSBr and $CrPS_4$. Their crystal structures, magnetic properties, and electronic band structures will be comprehensively described below.

2.1 Crystal structures

CrX3 (X = Cl, Br, I). Chromium trihalides are substances in which vdW interfaces between layers X-Cr-X are dominant [28]. Within each layer, the chromium trihalides are isostructural and consist of a honeycomb structure of edge-sharing octahedral, with central chromium coordinated to six monovalent halide anions at the corner, as shown in **Figure 1a** [24, 29, 30]. Besides, these layers stack in monoclinic phase at a high temperature with space group *C*2/m: $CrCl_3$ (above 240 K) [31–33], $CrBr_3$ (above 420 K) [33] and CrI_3 (above 220 K) [30]. On the other hand, these layers stack in rhombohedral phase with space group $R\overline{3}$ at a low temperature, as shown in **Figure 1b**. Thus, bulk CrX_3 crystals are in the low-temperature rhombohedral crystallographic phase at room temperature [25]. **MPS_3 (M = Fe, Ni, Mn).** The transition metal phosphorous trisulfides share a defining common structural feature such that $(P_2S_6)^{4-}$ anion sublattice appears within each layer crystal. The honeycomb arrangement of the transition metal ions is distributed around $(P_2S_6)^{4-}$ bipyramids, as show in **Figure 2a** [36–39]. **CrSBr** shows a layered crystal structure with vdW interactions, in which each layer is made of two buckled planes of CrS sandwiched between Br sheets and magnetic Cr^{3+} ions form a rectangular lattice, as shown in **Figure 3a** [40, 43, 44]. The crystal symmetry remains unchanged between 15 and 300 K [45]. **$CrPS_4$** possesses a monoclinic symmetry with C2/m space group, in which puckered layers of S atoms are in hexagonal close packing parallel to the *a*-axis and the Cr atoms form a square lattice [46, 47]. In each layer, the Cr atom is surrounded by six S atoms, resulting in this distorted CrS_6 octahedron. The P atom is coordinated in the center of three CrS_6 octahedra. The weak vdW force is between sulfur layers, as shown in **Figure 4a** [49–51].

2.2 Magnetic properties

CrX3 (X = Cl, Br, I) exhibit multiple magnetic phases. The superexchange coupling occurs between two adjacent Cr^{3+} sites and is mediated through the halide ions, which serves as the predominant exchange pathways in the chromium trihalides, as shown in **Figure 1c**. The interlayer exchange interaction between Cr atoms in different layers. **Figure 1d** presents the details of the magnetic order of atomically thin CrX_3. Within a single layer, the chromium moments of CrX_3 are ferromagnetically coupled as intuited by the Goodenough-Kanamori-Anderson rules [29, 52, 53]. In the case of monolayer CrI_3 and $CrBr_3$, the magnetic moments aligned perpendicular to the crystal plane [30, 54], while they aligned parallel to the crystal plane for monolayer $CrCl_3$ [33, 55–57]. In the bulk, two adjacent layers of CrI_3 and $CrBr_3$ show FM interlayer exchange [30, 54], and AFM interlayer exchange in $CrCl_3$ [33, 55–57]. However, for bilayer CrI_3, two adjacent layers show AFM interlayer exchange [17]. Moreover, the corresponding magnetic ordering temperatures are 61 K for bulk CrI_3 [30], 37 K for bulk $CrBr_3$ [58], and 17 K for bulk $CrCl_3$ [33]. **MPS3 (M = Fe, Ni, Mn)** are a class of vdW stacking AFM insulator, which possess different AFM ordering with the different transition metal due to their different spin dimensionalities. The magnetic ground states of MPS_3 are summarized in **Figure 2b**. All of the compounds are AFM, but only $MnPS_3$ acquires the Néel state where each magnetic site is anti-aligned with all nearest neighbors [59, 60]. In contrast, the ground states of $FePS_3$ and $NiPS_3$ exhibit Zigzag-type ordering, where each transition metal atom is aligned with two nearest neighbors and anti-aligned with one nearest neighbor [61, 62]. Based on the Mermin-Wagner theorem, $FePS_3$ is predicted to have an easy axis, whereas $NiPS_3$ and

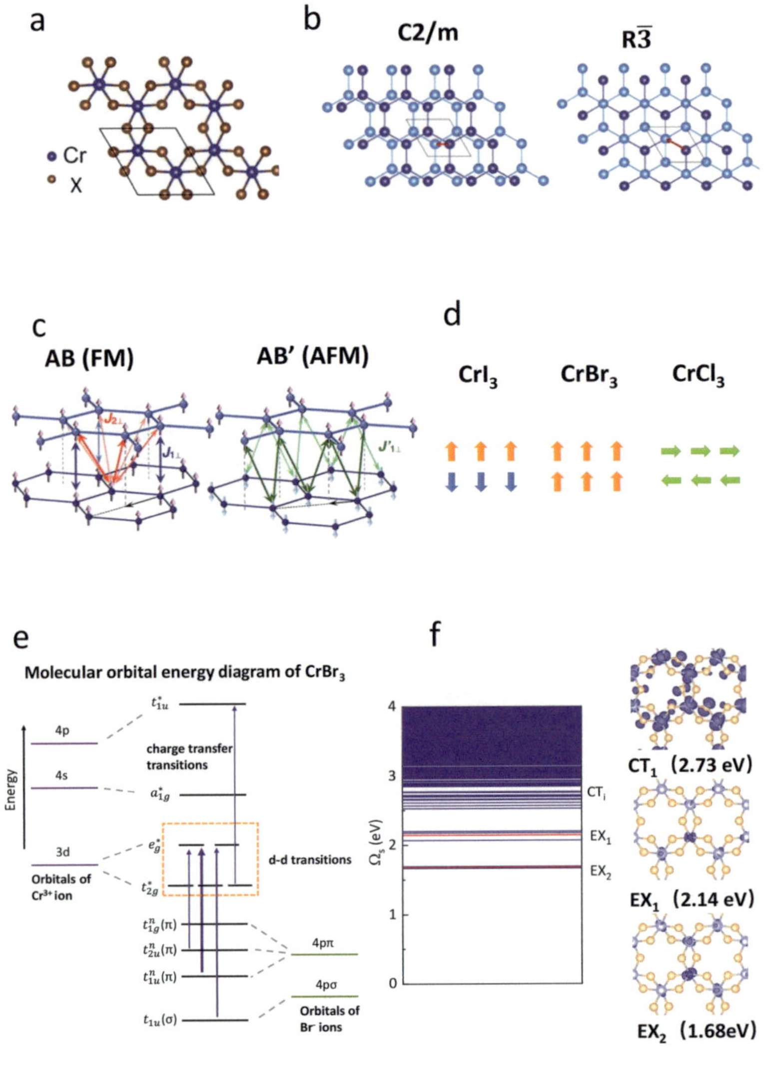

Figure 1.
a, Crystalline structure of CrX_3 in top view showing the honeycomb arrangement of the Cr atoms [24] Copyright 2022, American Physical Society. b, Top view of the CrX_3 layers in the monoclinic (left) and rhombohedral (right) phases. Two CrX_3 layers without the X– ions to show the relative stacking order of the bottom (dark blue) and top (blue) Cr^{3+} layers in two phases [25] Copyright 2021, American Chemical Society. c, Interlayer exchange of CrX_3: interlayer Cr nearest-neighbor ($J_{1\perp}$, in blue) and the second-neighbor ($J_{2\perp}$, in red) in AB-stacking (left) and the nearest-neighbor ($J'_{1\perp}$, in green) in AB'-stacking (right) [26] Copyright 2018, American Chemical Society. d, Illustrations showing the magnetic ground states of bilayer CrI_3 (left), $CrBr_3$ (middle), and $CrCl_3$ (right). e, Molecular orbital energy diagram for $CrBr_3$. The d-d LF transitions arise from the ground and excited configurations of the d^3 electrons in the t_{2g}^* and e_g^* orbitals. The dashed orange box shows d-d transitions and the blue arrows show possible LMCT transitions. f, Exciton energy levels of monolayer $CrBr_3$ calculated using the first-principles GW-BSE method (left). Top view of excitonic wave function in real space for selected states that are responsible for the exciton bands of CT_1, exciton1 (EX_1) and exciton2 (EX_2) (right) [27] Copyright 2022, American Chemical Society.

$MnPS_3$ both have easy planes coinciding with the atomic planes [62]. The Néel temperatures are 118 K for bulk $FePS_3$ [63–65], 155 K for bulk $NiPS_3$ [66], and 78 K for bulk $MnPS_3$ [61]. **CrSBr** is an AFM semiconductor with the A-type antiferromagnetism, which is ascribed to the halogen-mediated (Cr-Br-Cr) and chalcogen-mediated

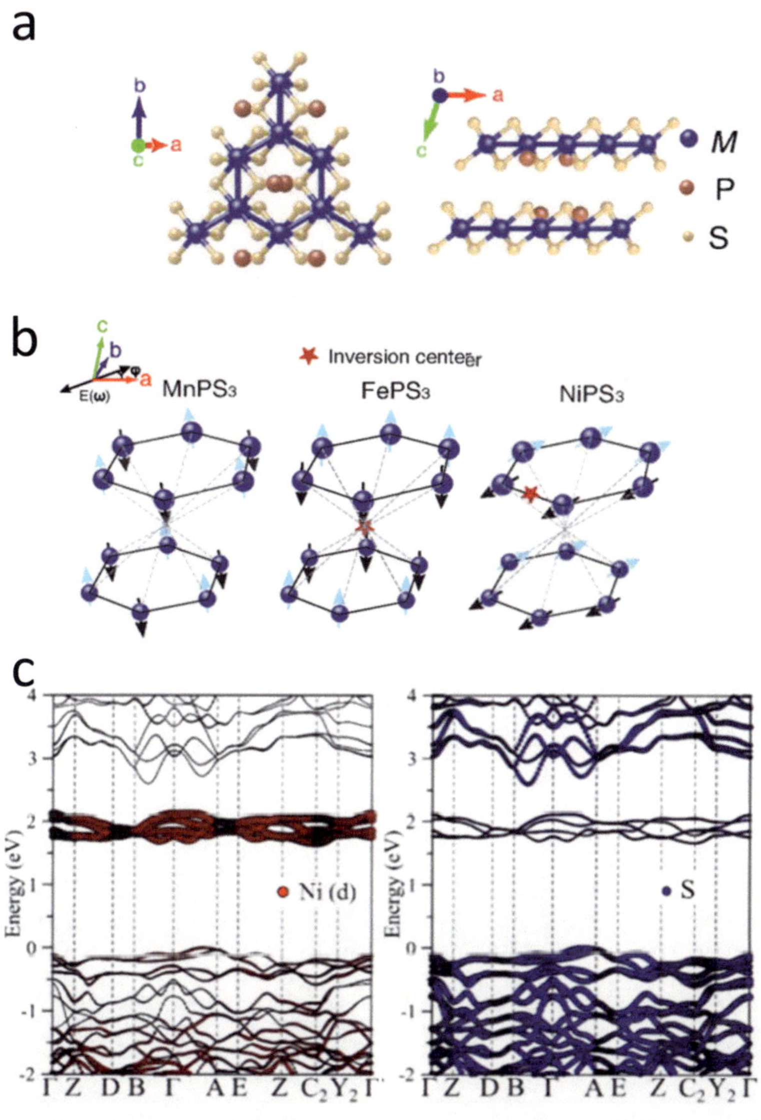

Figure 2.
a, Crystal structure of MPS_3 viewed along the c and b axis [34] Copyright 2020, American Physical Society. b, The magnetic ground states of bilayer $MnPS_3$, $FePS_3$ and $NiPS_3$. Arrows denote spin orientation. Star denotes an inversion center of the AFM structure [34] Copyright 2020, American Physical Society. c, Calculated electronic band structure of bulk $NiPS_3$ with the red and blue circles corresponding to projected bands from Ni d-orbitals and S atoms. [35] Copyright 2022, AAAS.

(Cr-S-Cr) strong superexchange interactions and weak interlayer coupling [40, 67]. As a result, each rectangular layer exhibits in-plane anisotropic FM order, and these FM layers couple antiferromagnetically along the stacking direction, as shown in **Figure 3b** [41, 43, 44]. The AFM Néel temperature is 132 K for bulk CrSBr and FM Curie temperature is 180 K [40, 43]. **CrPS4** is a promising vdW AFM semiconductor with the A-type antiferromagnetism, consisting of out-of-plane FM monolayers coupled antiferromagnetically, as shown in **Figure 4b** [48, 68]. The Néel temperature is ~36 K for bulk $CrPS_4$, below which the spin-flip transition from AFM to FM orders occurs [69, 70].

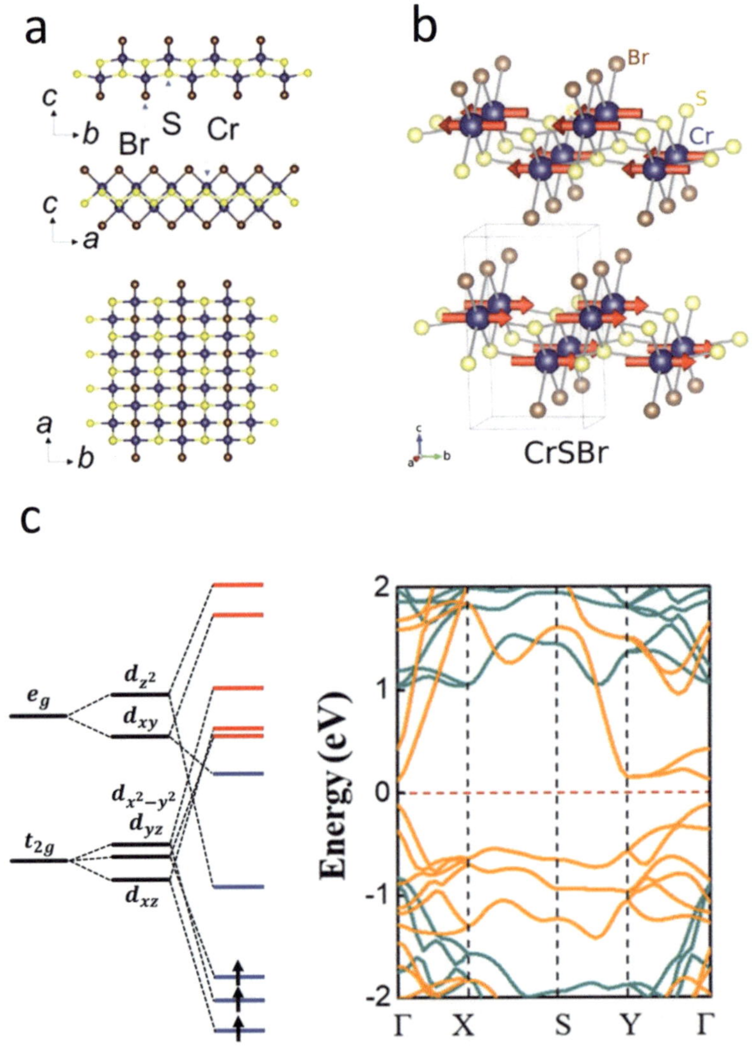

Figure 3.
a, Side and top views of the crystal structure of CrSBr [40] Copyright 2022, Wiley-VCH GmbH. b, The magnetic order of CrSBr, FM in-plane but layered in alternating directions for a bulk antiferromagnetism [41] Copyright 2022, Wiley-VCH GmbH. c, Splitting of d orbitals of the Cr atom under the octahedral crystal field of the CrSBr monolayer and the spin-polarized band structure of FM CrSBr [42] Copyright 2018, American Chemical Society.

2.3 Electronic band structures

CrX3 (X = Cl, Br, I). In this crystal field geometry, the Cr *d* splits into a t_{2g}^* triplet and an e_g* doublet. Cr^{3+} has a valence of three electrons, which fill the t_{2g} majority-spin band according to Hund's first rule, leaving all other *d* bands empty. Since all three CrX_3 compounds have FM order down to the monolayer below Curie temperatures, we use $CrBr_3$ as an example to present the molecular orbital energy diagram based on ligand-field (LF) theory, as shown in **Figure 1e**. In the dashed orange box allows *d-d* transitions and the blue arrows show possible ligand-to-metal charge transfer (LMCT) transitions [71]. However, an accurate first-principles calculation

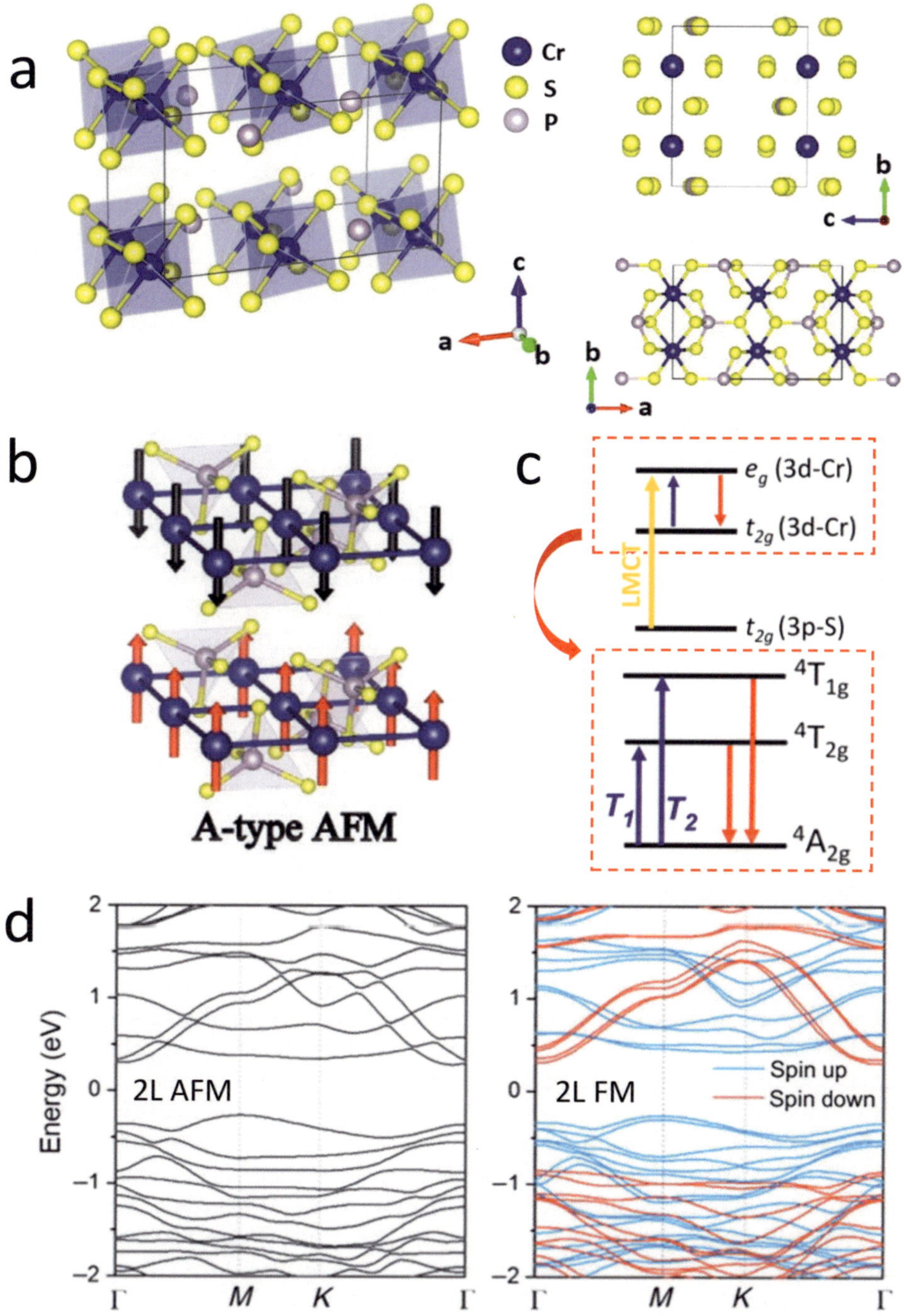

Figure 4.
a, Crystal structure of $CrPS_4$ with the C2/m space group [46] Copyright 2020, Springer Nature. b, Magnetic structure of $CrPS_4$, the back and red arrows point to the direction of magnetic moments [48] Copyright 2020, WILEY-VCH Verlag GmbH and Co. KGaA. c, Molecular orbital energy diagram in the optical transitions of $CrPS_4$ (top) and configuration diagram of Cr^{3+} for the d-d transitions (bottom) [46] Copyright 2020, Springer Nature. d, Spin-dependent band structures for bilayer $CrPS_4$ in the AFM and FM states [49] Copyright 2021, American Physical Society.

of the electronic structure of CrX_3 should account for both the dielectric polarization from the ligand groups and the on-site Coulomb interactions among the localized spin-polarized electrons in 2D limit [24, 27, 72]. Thus, the recent work has achieved the accurate electronic band structures of $CrBr_3$ monolayer with first-principles GW-BSE calculations including the excitonic effect, as shown in **Figure 1f** [27]. The calculation yields a series of bright exciton energy levels with energies of 1.68, 2.14, and 2.72 eV, which coincides with the measured absorption spectrum. The real-space exciton wave functions corresponding to LMCT, EX_1 and EX_2 states are shown in **Figure 1f** [27]. The electron and hole are localized separately on the anion and cation for LMCT state, which confirms the CT characteristics of higher energy absorptions in $CrBr_3$. While the excitonic wave functions predominantly occupy the Cr atoms for EX_1 and EX_2 states, confirming the *d-d* origin. **MPS3 (M = Fe, Ni, Mn).** Here, we mainly discuss the electronic band structure of $NiPS_3$, since the band-edge excitons have been observed in few-layer $NiPS_3$ with interesting optical phenomena. $NiPS_3$ exhibits a unique electronic structure with a highly localized electronic band composed by *d* orbitals. A recent work calculated the electronic band structure and density of states for bulk $NiPS_3$ using first-principles density functional theory + U method, as shown in **Figure 2c** [35, 39, 73, 74]. The electronic bands near the conduction band minimum are predominantly contributed by Ni *d* orbitals, the small dispersion of these bands is due to the localization of the Ni *d* orbitals. In contrast, the valence bands are mostly from S *p* orbitals. The calculation indicates an indirect band gap of ~1.6 eV, consistent with the previous results [75, 76]. However, the most recently published work demonstrates that Zhang-Rice triplet (ZRT) state and Zhang-Rice singlet (ZRS) state are formed from the spin-orbital coupling between Ni *p* orbital and S *p* orbitals in APM $NiPS_3$. The transition from ZRT to ZRS states will host a spin-orbit-entangled exciton state [77]. **CrSBr** combines a direct electronic band gap with layered A-type AFM order. Initially, due to the symmetry breaking under the octahedral crystal field, five degenerate *d* orbitals of the Cr atom are split into e_g levels ($d_z{}^2$ and d_{xy}) with higher energy and t_{2g} levels ($d_x{}^2$-$_y{}^2$, d_{yz}, and d_{xz}) with lower energy, as shown in **Figure 2c** (left) [42]. In the monolayer CrSBr structure, distortion of octahedron leads to further splitting of e_g and t_{2g} levels, consequently, the five d orbitals are no longer degenerate. With the coupling with atomic orbitals of S atoms, the spin-polarized band structures of CrSBr are formed, as shown in **Figure 2c** (right) [42]. Furthermore, the more accurate electronic structure calculations of monolayer CrSBr in its FM ground state within the GW approximation demonstrate a semiconducting band gap of ~1.8 eV and highly anisotropic band dispersion [78]. First-principles GW-BSE calculations unveil the different exciton wave functions for AFM and FM CrBrS bilayers. In the AFM bilayer, the electron is localized in the same layer as the hole. In the FM bilayer, the electron wave function is delocalized across both layers for a hole fixed in the bottom layer [78]. Thus, the optical transitions dominant by these excitonic transitions will be tuned during the AFM to FM phase transition. According to the crystal structure of **CrPS4**, each Cr atom is located in a distorted octahedral interstice formed by six S atoms, thus inducing an octahedral ligand perturbation field that splits that 3*d* orbitals of Cr^{3+} into t_{2g} and e_g orbitals. The ground state ($^4A_{2g}$) and three lowest excited states (^2E_g, $^4T_{2g}$, $^4T_{1g}$) emerge from the ground term (^4F) of d^3, as shown in **Figure 4c** [46, 79]. The optical transitions should be from the spin-allowed *d-d* transition of the Cr^{3+} ion. For more realistic scenarios, spin is considered in the electronic band structure calculations, as shown in **Figure 4d**, the band gap for bilayer $CrPS_4$ is 0.83 eV (AFM state), 0.72 eV (spin up in the FM state) and 1.43 eV (spin down in the FM state), respectively [49]. However, for more accurate electronic band structures of $CrPS_4$, the excitonic effect should be considered.

3. Light-mater interactions in 2D magnets

vdWs layered materials offer fascinating opportunities for studying light-matter interactions in the 2D limit. So far, many interesting optical phenomena have been observed in monolayer semiconducting TMDs and other non-metallic 2D materials. However, none of these 2D materials possesses long-range magnetic order. The discovery of 2D system hosting intrinsic magnetic order represents a seminal addition to the rich landscape of vdW materials. More importantly, the interdependence of structure and magnetism, along with strong light-matter interactions, provides a new platform to explore the optical control of magnetic, electronic, and vibrational degrees of freedom at the 2D limit. Here, we briefly summarize the optical properties of several 2D magnets that have been studied in recent years. The correlation between emitted photons, phonons and spins (magnetic orders) in layered magnets provides routes for investigating magneto-optics in 2D materials, and hence opens a path for developing opto-spintronic devices and AFM/FM-based quantum information technologies.

3.1 Light absorption and emission of 2D magnets

The excitonic absorption and emission properties have been widely studied in **CrX3 (X = Cl, Br, I)**, which can be tuned by changing their magnetic ordering. **Figure 5a** presents the optical absorption and photoluminescence (PL) properties of bulk $CrBr_3$ in FM state (10 K) [27]. It is clear that two excitonic absorption peak X_1 and X_2 are located in the visible region, and PL is in the near-infrared region. As explained in this work PL originates from the *d*-*d* transitions modified by polaronic effect in $CrBr_3$, resulting in a ultra long emission lifetime of around 4.3 μs [27, 80]. Besides, the *d*-*d* transition in the out-of-plane magnetic field shows circularly selective PL [71]. **Figure 5b** shows the PL from monolayer $CrBr_3$ at 2.5 K in −0.5 and 0.5 T with $\sigma^+\sigma^+$ and $\sigma-\sigma-$ excitation-collection configurations, where $\sigma^+(\sigma-)$ represents the left (right) circularly polarized light. The PL at ±0.5 T shows opposite helicity, indicating that the spin of electrons in $CrBr_3$ is coupled to the circularly polarized light [80]. **NiPS3** exhibits strongly correlated electrons, where the excitons' unique coupling to spin and orbital degrees of freedom in this AFM 2D magnet. The non-equilibrium driven of such dressed quasiparticles offers a promising platform for realizing unconventional many-body phenomena and phased beyond thermo-dynamic equilibrium. Recently, spin-orbital-entangled excitons that arise from Zhang-Rice state have been achieved and observed in this vdW correlated insulator $NiPS_3$ under photo excitation. As shown in **Figure 5c** (left), $NiPS_3$ has a charge-transfer (CT) gap of ~1.8 eV at 20 K (AFM state) [81]. Below the charge excitation lies a rich spectrum of sub-gap absorption resonances, including on-site *d*-*d* transitions at around 1.1 and 1.7 eV [81]. More importantly, there is a complex of spin-orbit-entangled excitons at around 1.5 eV, which has been assigned to the transition from a ZRT to ZRS state [77]. Besides, PL spectrum of a few-layer $NiPS_3$ sample with a thickness of about 10 nm shows two broadened emission peaks at

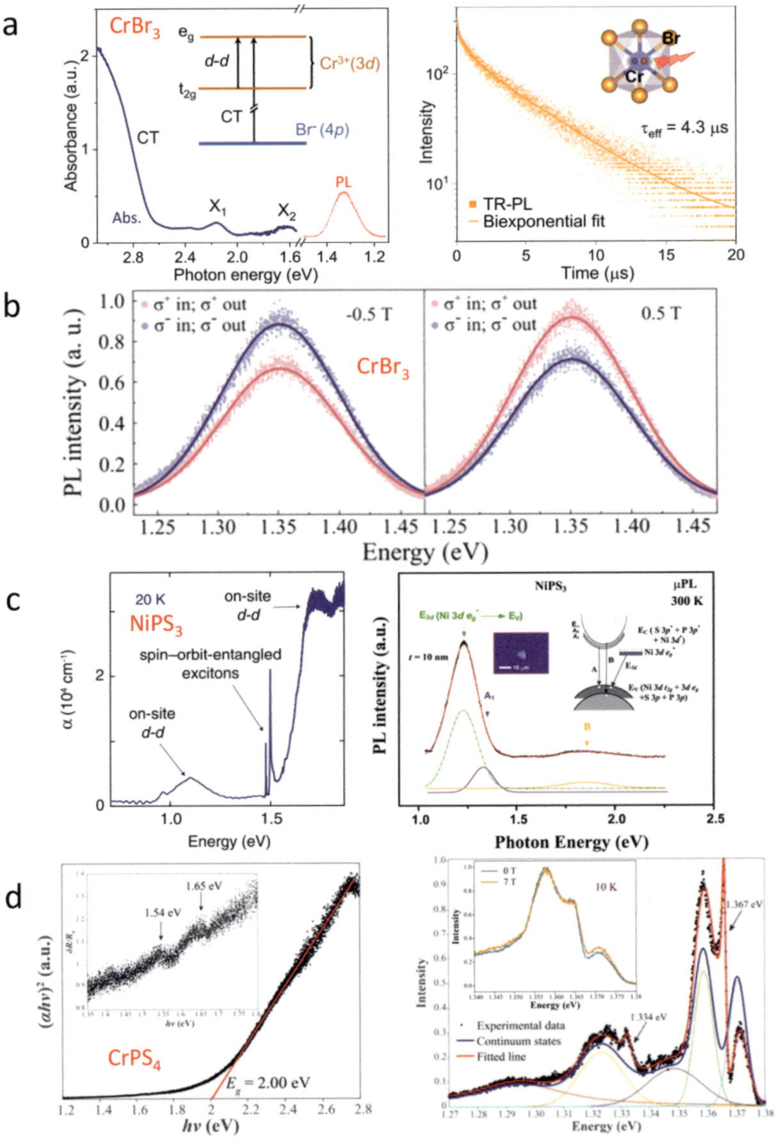

Figure 5.
a, Optical absorption and overlaid PL spectrum of $CrBr_3$ thick samples measured at 10 K (left). The d-d (X_1 and X_2) and CT transitions are marked and illustrated in a simplified energy diagram in the inset. Time-resolved PL decay curve obtained at 4 K by integrating the PL intensity over the spectral range of 1.25–1.4 eV (right). The fitting (solid line) yields an averaged lifetime of 4.3 μs. Inset: schematic of light emission from Cr^{3+} metal ion in the octahedral crystal field [27] Copyright 2022, American Chemical Society. b, Spontaneous circularly polarized PL spectra for $\sigma^+\sigma^+$ (red) and σ–σ– (blue) circular polarization components from monolayer $CrBr_3$ at 2.7 K under the applied out-plane magnetic field of - 0.5 T (left) and 0.5 T (right) [80] Copyright 2019, American Chemical Society. c, Optical absorption (α) of $NiPS_3$ below the CT gap at 20 K (left). The features around 1.5 eV at the spin-orbital-entangled exciton transitions. The broad structures around 1.1 and 1.7 eV are on-site d-d transitions [81] Copyright 2021, Springer Nature. PL spectra of the band-edge emissions of a few-layer $NiPS_3$ sample at 300 K (right) [82]. Copyright 2021, Springer Nature. The fitting curves of the PL peaks are also included. The left inset shows a microscopic image of the few-layer sample, and the right inset shows the possible band-edge scheme for the observed PL emissions. d, Differential reflectance spectrum of a few-layer $CrPS_4$ flake (black dots), and the liner fitting (red line) reveals the absorption edge at 2.00 eV (left). Inset: the zoomed-in spectrum. PL spectrum of $CrPS_4$ at 4 K fitted by the Fano formula (right). The continuum band is composed of four Gaussian peaks and one Lorentzian peak. Inset: PL spectra obtained at 10 K with applied field of 0 T and 7 T [51] Copyright 2020, American Chemical Society.

E_{3d} ~1.23 eV and at B ~1.825 eV at 300 K, as shown in **Figure 5c** (right). A detailed analysis of the PL line-shape fitting of the PL spectrum reveals that the A_1 peak (~1.366 eV) is still involved in the broadened E_{3d} peak [82]. Further thickness-dependent PL intensity results indicate E_{3d} peak may be from the Ni 3*d* e_g^* band to the E_v transition assisted by phonons at 300 K. The inset of **Figure 5c** (right) shows the probable band scheme of the band edge for PL band-edge emissions. The E_{3d} peak is an indirect-like emission caused by the intermediate band of Ni 3*d* e_g^* existed in $NiPS_3$. The A_1 and B emissions are originated from the direct E_c bottom to the E_v top and to the spin-split-off band below E_v, respectively [82]. **$CrPS_4$** is a promising ternary AFM semiconductor with PL in the near-infrared wavelength region. Recently, a Fano resonance, arising from quantum interference between a discrete optical transition and a continuous background, is observed in PL from $CrPS_4$ flakes with decreasing temperature below the Néel temperature. **Figure 5d** (left) shows the plot of $(\alpha h\nu)^2$ versus the photon energy $h\nu$, which can determine the absorption edge is ~2.00 eV ascribed to the CT transitions from the 3*p* of the S band to unoccupied 3*d* band of the Cr atom. In the energy below the absorption edge, two weak peaks can be observed at 1.54 and 1.65 eV, as show in the inset of **Figure 5d** (left) [51]. At the low temperature 4 K, PL emission in the near-infrared region is observed, where PL shows several narrow peaks with two asymmetric profiles located at 1.333 and 1.367 eV emerging [79]. Such an asymmetric line shape is a typical feature of the Fano resonance, which can be well described by a Fano formula (the fitted red solid line). This work explains that the continuous background responsible for the Fano resonances is attributed to the *d-d* transition of the Cr^{3+} Center, predominantly the spin-forbidden ^2E_g to $^4A_{2g}$ transition with contributions of the broad-band $^4T_{2g}$ to $^4A_{2g}$ transition [51]. Since this Fano resonance is only observed with the temperature below the Néel temperature, it indicates that AFM order of $CrPS_4$ is of great importance to the occurrence of this quantum inference. However, the applied out-of-plane magnetic field shows no obvious influence on the PL spectra, as demonstrated by the inset of **Figure 5d** (right), which may be due to the saturation of the discrete state transition at 10 K.

3.2 Light scattering of 2D magnets

Raman scattering measures light in elastically scattered from collective quasiparticle excitations. In particular, Raman scattering from spin-phonon coupling, electron-phonon coupling, phase transitions, and magnetic excitations has yielded incisive information on recently developed 2D magnets. **CrI3** has been demonstrated that hosts various interesting light scattering properties: (i) the tuning of in elastically scattered light through symmetry control in atomically thin CrI_3, as shown in **Figure 6a** [83, 87]. Raman spectra taken at 5 K in the cross-linear polarization channel (xy) shows a forbidden Raman activity of A_g series mode, as shown in **Figure 6a** (left) [83]. In contrast, there are two modes appearing in the cross-linearly polarized Raman scattering at 77 and 126 cm^{-1}, labeled as P_1 and P_2 in **Figure 6a** (right) [83]. These modes have been previously attributed to one-magnon excitations since they appear in the magnetically ordered state and have their largest intensity in xy configuration, as shown in the inset of **Figure 6a** (right). Besides, in bilayer CrI_3, the A_{1g} phonon mode becomes Davydov-split into two modes of opposite parity, which exhibit divergent selection rules that depend on inversion symmetry and the underlying magnetic order. At zero magnetic field (AFM state), there are two peaks distinct in energy appeared at 126.7 and 128.8 cm^{-1} in the cross- and co-linearly

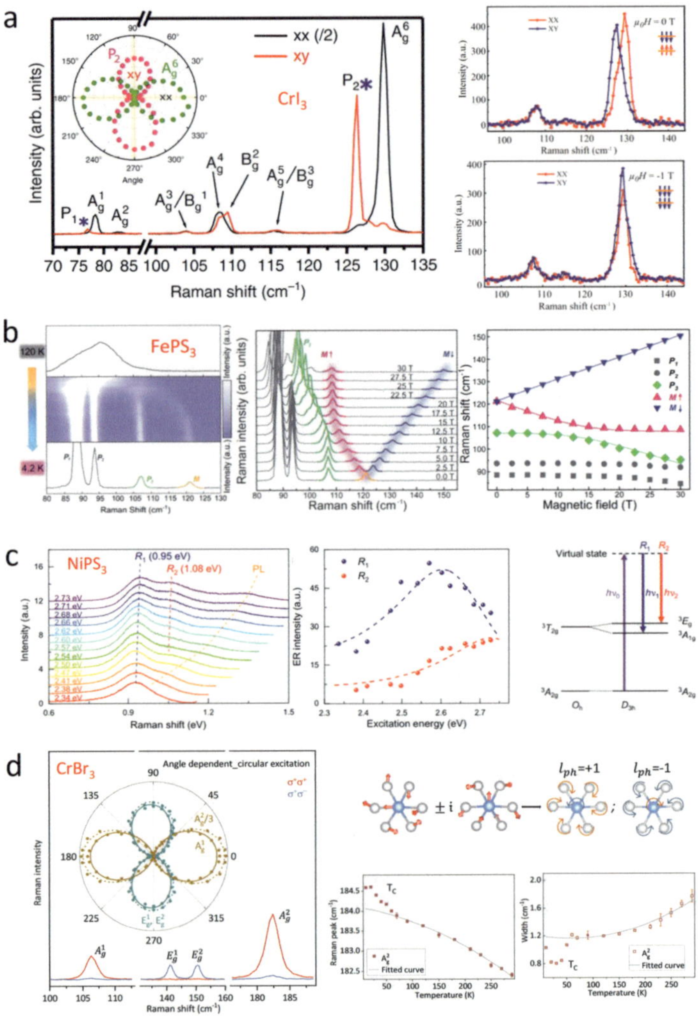

Figure 6.

a, The left: Raman spectra of few-layer CrI_3 at 5 K (0 T) for xx (black) and xy (red) polarization configurations. Spectra in xx were divided by two for clarity. Peaks that appear below T_c are highlighted with blue asterisks. Inset shows intensity as a function of collection polarization angle for P_2 and A_g^6 in a polar plot [83] Copyright 2020, Springer Nature. The right: co- and cross-linearly polarized Raman spectra taken in an AFM state at 0 T (top) and the fully spin-up polarized state at −1.0 T of A_g Raman mode from bilayer CrI_3 [84] Copyright 2020, American Chemical Society. b, Temperature-dependent Raman spectra of $FePS_3$ single crystal (left). A single broad peak at 120 K splits into four sharp peaks at 4.2 K. Raman spectra of $FePS_3$ in magnetic field 0–30 T (middle). Raman spectra have been vertically shifted for clarity, in steps of 2.5 T. Zeeman splitting of the M branch identifies it as magnon mode. The two branches are denoted as $M_\uparrow$ (red) and $M_\downarrow$ (blue). Magnetic-field-dependent peak position of P_1, P_2, P_3, $M_\uparrow$ and $M_\downarrow$, which are denoted as black squares, black circles, green diamonds, red up-triangles, and blue down-triangles (right). The solid lines are fitted to the data [85] Copyright 2021, American Physical Society. c, Optical spectra of $NiPS_3$ plotted in terms of Raman shift, i.e., energy difference from excitation energy, excited by different lasers with energies from 2.73 to 2.34 eV (left). Integrated intensity of two electronic Raman (ER) peaks as a function of excitation energy (middle). The dashed lines are the fitting curves. Schematic illustration of ER scattering in between the ground state and first excited triplet state for Ni^{2+} ion in a trigonally distorted octahedral environment (D_{3h}) (right). O_h represents the crystal field splitting in an octahedral field; $h\nu_0$ and $h\nu_1$ ($h\nu_2$) are the energy of incident and scattered photons of R_1 (R_2) mode [35] Copyright 2022, AAAS. d, Helicity-resolved Raman spectra of bulk $CrBr_3$ at 10 K (left). Inset: polar plot of the Raman intensities of these four modes versus the rotation of the quarter-wave plate. The 0° and 90° correspond to the ($\sigma^+\sigma^+$) and ($\sigma^+\sigma^-$) configuration, respectively. Right: superposition of two orthogonal linear vibrations of E_g^1 (143.0 cm^{-1}) results in right-handed or left-handed circular motions at the Γ-point, i.e., chiral phonons with PAM of $l = \pm 1$ [86] Copyright 2021, Wiley-VCH GmbH.

polarized channels, respectively (top right corner of **Figure 6a**) [84]. When the field was above the spin-flip transition (−1 T) to fully align the spins into FM states, the 126.7 cm^{-1} peak presented in the AFM states is abruptly suppressed and only 128.8 cm^{-1} peak is observed in both polarized channels (bottom right corner of **Figure 6a**) [84]. These findings shed light on exploring the emergent magneto-optical effects in 2D magnets. (ii) the 2D magnons has been directly observed in atomically thin CrI_3 [88]. The ultra-low frequency magneto-Raman spectroscopy probed the effects of symmetry on magnetic excitations in monolayer CrI_3, a low-frequency feature emerges on both Stokes and anti-Stokes sides when a magnetic field is applied at −4 T. Interestingly, these Stokes and anti-Stokes modes obey opposite optical selection rules: for a given magnetization orientation, Stokes mode only appears when excited by on helicity of light, while anti-Stokes mode emerges with excitation of the opposite helicity. Further magnetic-field and temperature dependence measurements demonstrate that these Stokes and anti-Stokes modes are the signature of an acoustic magnon. Besides, strong coupling between magnons and phonons are directly observed in a 2D AFM semiconductor $FePS_3$, vis magneto-Raman spectroscopy at magnetic fields up to 30 T [85]. The temperature-dependent Raman spectra on $FePS_3$ shows a Raman active magnon peak at 121 cm^{-1}. And the broad Raman band evolves into three sharp Lorentz peaks, P_1, P_2 and P_3 as temperature decreased to below T_N, as shown in **Figure 6b** (left). The evolution of the Raman spectrum of $FePS_3$ in an out-of-plane magnetic field shows that P_1, P_2 and P_3 peaks have no energy shift with low magnetic field, therefore verified as phonons, as shown in **Figure 6b** (middle). In contrast, the peak M exhibits a Zeeman splitting, confirming its magnonic character. These two split peak sections are labeled as $M_\downarrow$ and $M_\uparrow$, which have opposite spins. The $M_\downarrow$ and $M_\uparrow$ peak positions of show linear dependence on a magnetic field with similar slopes (0.99 and −0.94 cm^{-1}/T), which is the effective g factor of the magnon, as shown in **Figure 6b** (right). Until 22.5 T, the shift of P_1 and P_2 is negligible, while the field-driven anticrossing of $M_\uparrow$ and P_3 with a 6.1 cm^{-1} energy gap signals a repulsive interaction between magnon $M_\uparrow$ and P_3 modes when their energies are in resonances, which is evidence that two modes are strongly coupled [85]. Due to the correlated-electron system of $NiPS_3$, the scattering of incident photons with d electrons in Ni^{2+} ions has been observed at ~1.0 eV [35]. As shown in **Figure 6c** (left), the measured distinct optical spectra as the excitation wavelength varying from 454 nm (2.73 eV) to 531 nm (2.34 eV), show a peak at the same position ~1.39 eV assigned to the phonon sideband of the coherent ZR exciton. In contrast, there are two other peaks, R_1 and R_2 exhibiting an obvious redshift with decreasing of the excitation energy, and, finally, disappear when the excitation energy is below ~2.35 eV. These two peaks present a fixed energy difference between the collected signal and the laser excitation, suggesting that their origin is Raman scattering rather than luminescence. The intensity of electronic Raman peaks of R_1 and R_2 exhibit a clear dependence on the excitation energy, indicating a resonant effect, which is related to an external electronic state as an intermediate state, as shown in **Figure 6c** (right). Fitting results demonstrate that the intensities of R_1 and R_2 reach the maximum at ~2.61 and ~2.73 eV, respectively. The different resonant energies of these two modes indicate a scattered light resonance. After calculation, they obtained the intermediate state energy of ~1.70 eV, which matches the energy of the second excited states ($^3T_{1g}$) from the ground states. The electronic Raman scattering process is proposed in between the ground state and first excited triplet state for Ni^{2+} ion in a trigonally distorted octahedral environment, as shown in **Figure 6c** (right). Phonon chirality and spin-phonon coupling have been observed in 2D **CrBr3**

[86]. The helicity-resolved Raman scattering measurement at 10 K can be used to resolve the chirality of phonon modes in $CrBr_3$ (FM phase), since chiral phonon modes originate from the circular vibration of sublattices. As shown in **Figure 6d** (left), the Raman peaks corresponding to E_g modes appear only in the cross-circularly polarized configuration ($\sigma^+\sigma^-$), which means that Raman scattering involved E_g modes reverses the helicity of the incident photons. In contrast, Raman scattering involving the two A_g modes does not change the helicity of the incident photons. The corresponding polar plot of E_g and A_g modes clearly presents their distinct polarization properties, where E_g modes have maximum Raman intensity in the cross-circularly polarized configuration ($\sigma^+\sigma^-$) rather than the almost extinct A_g modes (inset in the left side of **Figure 6d**). Since Raman scattering involving E_g modes switches the polarization of circularly polarized light, the E_g modes in $CrBr_3$ must have a non-zero pseudo-angular momentum (PAM), based on the conservation of PAM in the Raman scattering process. Besides, single-phonon Raman scattering process can only induce zone-center phonons due to the conservation of crystal momentum. However, the zone-center phonons have real eigenvectors which have zero phonon circular polarization by definition. Thus, the only zone-center Raman active phonon modes that can reverse the helicity of incoming light are complex superposition of the degenerate E_g modes, as demonstrated by the top right corner of **Figure 6d**. Furthermore, it is nature to explore the magnetic effect on the phonon properties of $CrBr_3$ upon cooling, given the existence of a 2D long-range FM order below Curie temperature in $CrBr_3$. As shown in **Figure 6d** (the bottom right corner), the Raman peak shift and linewidth change of A_g mode with decreasing temperature exhibit an anomalous behavior when the temperature is near to the Curie temperature. For example, in the temperature range between 290 and 50 K, a conventional hardening of phonon with decreasing temperature is observed due to the suppression of the anharmonic phonon-phonon interactions. While such temperature dependence becomes much stronger with the onset of magnetic ordering below ~50 K, as demonstrated by the additional anomalous phonon hardening due to the short-range local ordering of magnetic moments.

4. Conclusion

In summary, the recently developed 2D magnets and their novel photo-physical phenomena have been briefly summarized in this chapter. Firstly, we introduced the crystal structures, magnetic properties, and electronic band structures of several hot 2D magnets like CrX_3 (X = Cl, Br, I), MPS_3 (M = Fe, Ni, Mn), CrSBr and $CrPS_4$. Then, according to these features, they exhibited novel light absorption, emission and scattering properties. More importantly, 2D magnets provide us a platform to tune the magnetic, electronic, and vibrational degrees of freedom for designing novel 2D magnet-based device applications. Although great processes have been made in this booming field in the past few years, the research on exploring light-matter interactions in 2D magnets is still in its infancy. There are many aspects that need researchers to further explore: growth of new 2D magnets, fabrication 2D magnetic heterostructures, development of advanced optical characterization techniques, and applying various state-of-the-art external perturbations for effectively engineering, et al. In essence, 2D magnets offer us an entirely new opportunity to explore novel photo-physical phenomena, therein looking promising for many more exciting outcomes shortly.

Acknowledgements

T. Yin gratefully acknowledges strong support from the Presidential Postdoctoral Fellowship of Nanyang Technological University.

Conflict of interest

The authors declare no conflict of interest.

Author details

Tingting Yin
Division of Physics and Applied Physics, School of Physical and Mathematical Sciences, Nanyang Technological University, Singapore, Singapore

*Address all correspondence to: ttyin@ntu.edu.sg

References

[1] Novoselov KS, Geim AK, Morozov SV, Jiang D, Zhang Y, Dubonos SV, et al. Electric field effect in atomically thin carbon films. Science. 2004;**306**:666-669. DOI: 10.1126/science.1102896

[2] Geim AK, Novoselov KS. The rise of graphene. Nature Materials. 2007;**6**:183-191. DOI: 10.1038/nmat1849

[3] Mas-Ballesté R, Gómez-Navarro C, Gómez-Herrero J, Zamora F. 2D materials: To graphene and beyond. Nanoscale. 2011;**3**:20-30. DOI: 10.1039/C0NR00323A

[4] Novoselov KS, Jiang D, Schedin F, Booth TJ, Khotkevich VV, Morozov SV, et al. Two-dimensional atomic crystals. Proceedings of National Academy of Sciences of United States of America. 2005;**102**:10451-10453. DOI: 10.1073/pnas.0502848102

[5] Bhimanapati GR, Lin Z, Meunier V, Jung Y, Cha J, Das S, et al. Recent advances in two-dimensional materials beyond graphene. ACS Nano. 2015;**9**:11509-11539. DOI: 10.1021/acsnano.5b05556

[6] Mueller T, Malic E. Exciton physics and device application of two-dimensional transition metal dichalcogenide semiconductors. npj 2D Materials and Applications. 2018;**2**:29. DOI: 10.1038/s41699-018-0074-2

[7] Nikam RD, Lee J, Choi W, Kim D, Hwang H. On-chip integrated atomically thin 2D material heater as a training accelerator for an electrochemical random-access memory synapse for neuromorphic computing application. ACS Nano. 2022;**16**:12214-12225. DOI: 10.1021/acsnano.2c02913

[8] Lee PA, Nagaosa N, Wen X-G. Doping a Mott insulator: Physics of high-temperature superconductivity. Reviews of Modern Physics. 2006;**78**:17-85. DOI: 10.1103/RevModPhys.78.17

[9] Onsager L. Crystal statistics. I. a two-dimensional model with an order-disorder transition. Physics Review. 1944;**65**:117-149. DOI: 10.1103/PhysRev.65.117

[10] Arnold CS, Dunlavy M, Venus D. Magnetic susceptibility measurements of ultrathin films using the surface magneto-optic Kerr effect: Optimization of the signal-to-noise ratio. The Review of Scientific Instruments. 1997;**68**:4212-4216. DOI: 10.1063/1.1148368

[11] Hellman F, Hoffmann A, Tserkovnyak Y, Beach GSD, Fullerton EE, Leighton C, et al. Interface-induced phenomena in magnetism. Reviews of Modern Physics. 2017;**89**:025006. DOI: 10.1103/RevModPhys.89.025006

[12] Yang G, Wang R, Ge M, Guo M, Wang J, Ma R, et al. Switchable electronic and enhanced magnetic properties of CrI_3 edges. Physical Chemistry Chemical Physics. 2021;**23**:10518-10523. DOI: 10.1039/D0CP06155G

[13] Ostwal V, Shen T, Appenzeller J. Efficient spin-orbit torque switching of the semiconducting van der Waals ferromagnet $Cr_2Ge_2Te_6$. Advanced Materials. 2020;**32**:1906021. DOI: 10.1002/adma.201906021

[14] Wang Z, Gibertini M, Dumcenco D, Taniguchi T, Watanabe K, Giannini E, et al. Determining the phase diagram of atomically thin layered antiferromagnet $CrCl_3$. Nature Nanotechnology. 2019;**14**:1116-1122. DOI: 10.1038/s41565-019-0565-0

[15] Mermin ND, Wagner H. Absence of ferromagnetism or antiferromagnetism in one- or two-dimensional isotropic heisenberg models. Physical Review Letters. 1966;**17**:1133-1136. DOI: 10.1103/PhysRevLett.17.1133

[16] Huang F, Kief MT, Mankey GJ, Willis RF. Magnetism in the few-monolayers limit: A surface magneto-optic Kerr-effect study of the magnetic behavior of ultrathin films of Co, Ni, and Co-Ni alloys on Cu(100) and Cu(111). Physical Review B. 1994;**49**:3962-3971. DOI: 10.1103/PhysRevB.49.3962

[17] Huang B, Clark G, Navarro-Moratalla E, Klein DR, Cheng R, Seyler KL, et al. Layer-dependent ferromagnetism in a van der Waals crystal down to the monolayer limit. Nature. 2017;**546**:270-273. DOI: 10.1038/nature22391

[18] Gong C, Li L, Li Z, Ji H, Stern A, Xia Y, et al. Discovery of intrinsic ferromagnetism in two-dimensional van der Waals crystals. Nature. 2017;**546**:265-269. DOI: 10.1038/nature22060

[19] Burch KS, Mandrus D, Park J-G. Magnetism in two-dimensional van der Waals materials. Nature. 2018;**563**:47-52. DOI: 10.1038/s41586-018-0631-z

[20] Ningrum VP, Liu B, Wang W, Yin Y, Cao Y, Zha C, et al. Recent advances in two-dimensional magnets: Physics and devices towards spintronic applications. Research. 2020;**2020**:1768918. DOI: 10.34133/2020/1768918

[21] Li Q, Yang M, Gong C, Chopdekar RV, N'Diaye AT, Turner J, et al. Patterning-induced ferromagnetism of Fe_3GeTe_2 van der Waals materials beyond room temperature. Nano Letters. 2018;**18**:5974-5980. DOI: 10.1021/acs.nanolett.8b02806

[22] Gong C, Zhang X. Two-dimensional magnetic crystals and emergent heterostructure devices. Science. 2019;**363**:eaav4450. DOI: 10.1126/science.aav4450

[23] Thiel L, Wang Z, Tschudin MA, Rohner D, Gutiérrez-Lezama I, Ubrig N, et al. Probing magnetism in 2D materials at the nanoscale with single-spin microscopy. Science. 2019;**364**:973-976. DOI: 10.1126/science.aav6926

[24] Wu M, Li Z, Louie SG. Optical and magneto-optical properties of ferromagnetic monolayer $CrBr_3$: A first-principles GW and GW plus Bethe-Salpeter equation study. Physical Review Materials. 2022;**6**:014008. DOI: 10.1103/PhysRevMaterials.6.014008

[25] Wang D, Sanyal B. Systematic study of monolayer to trilayer CrI_3: Stacking sequence dependence of electronic structure and magnetism. Journal of Physical Chemistry C. 2021;**125**:18467-18473. DOI: 10.1021/acs.jpcc.1c04311

[26] Sivadas N, Okamoto S, Xu X, Fennie CJ, Xiao D. Stacking-dependent magnetism in bilayer CrI_3. Nano Letters. 2018;**18**:7658-7664. DOI: 10.1021/acs.nanolett.8b03321

[27] Yin T, You J-Y, Huang Y, Thu Do HT, Prosnikov MA, Zhao W, et al. Signature of ultrafast formation and annihilation of polaronic states in a layered ferromagnet. Nano Letters. 2022;**22**:7784-7790. DOI: 10.1021/acs.nanolett.2c01771

[28] Wang H, Eyert V, Schwingenschlögl U. Electronic structure and magnetic ordering of the semiconducting chromium trihalides $CrCl_3$, $CrBr_3$, and CrI_3. Journal of Physics: Condensed Matter. 2011;**23**:116003. DOI: 10.1088/0953-8984/23/11/116003

[29] Zhang W-B, Qu Q, Zhu P, Lam C-H. Robust intrinsic ferromagnetism

and half semiconductivity in stable two-dimensional single-layer chromium trihalides. Journal of Materials Chemistry C. 2015;**3**:12457-12468. DOI: 10.1039/C5TC02840J

[30] McGuire MA, Dixit H, Cooper VR, Sales BC. Coupling of crystal structure and magnetism in the layered, ferromagnetic insulator CrI_3. Chemistry of Materials. 2015;**27**:612-620. DOI: 10.1021/cm504242t

[31] Klein DR, MacNeill D, Song Q, Larson DT, Fang S, Xu M, et al. Enhancement of interlayer exchange in an ultrathin two-dimensional magnet. Nature Physics. 2019;**15**:1255-1260. DOI: 10.1038/s41567-019-0651-0

[32] Morosin B, Narath A. X-Ray diffraction and nuclear quadrupole resonance studies of chromium trichloride. The Journal of Chemical Physics. 2004;**40**:1958-1967. DOI: 10.1063/1.1725428

[33] McGuire MA, Clark G, Kc S, Chance WM, Jellison GE, Cooper VR, et al. Magnetic behavior and spin-lattice coupling in cleavable van der Waals layered $CrCl_3$ crystals. Physical Review Materials. 2017;**1**:014001. DOI: 10.1103/PhysRevMaterials.1.014001

[34] Chu H, Roh CJ, Island JO, Li C, Lee S, Chen J, et al. Linear magnetoelectric phase in ultrathin $MnPS_3$ probed by optical second harmonic generation. Physical Review Letters. 2020;**124**:027601. DOI: 10.1103/PhysRevLett.124.027601

[35] Wang X, Cao J, Li H, Lu Z, Cohen A, Haldar A, et al. Electronic Raman scattering in the 2D antiferromagnet $NiPS_3$. Science Advances. 2022;**8**:eabl7707. DOI: 10.1126/sciadv.abl7707

[36] Lee J-U, Lee S, Ryoo JH, Kang S, Kim TY, Kim P, et al. Ising-type magnetic ordering in atomically thin $FePS_3$. Nano Letters. 2016;**16**:7433-7438. DOI: 10.1021/acs.nanolett.6b03052

[37] Wang Y-M, Zhang J-F, Li C-H, Ma X-L, Ji J-T, Jin F, et al. Raman scattering study of magnetic layered MPS_3 crystals (M = Mn, Fe, Ni)*. Chinese Physics B. 2019;**28**:056301. DOI: 10.1088/1674-1056/28/5/056301

[38] Brec R. Review on structural and chemical properties of transition metal phosphorous trisulfides MPS_3. Solid State Ionics. 1986;**22**:3-30. DOI: 10.1016/0167-2738(86)90055-X

[39] Kim K, Lim SY, Lee J-U, Lee S, Kim TY, Park K, et al. Suppression of magnetic ordering in XXZ-type antiferromagnetic monolayer $NiPS_3$. Nature Communications. 2019;**10**:345. DOI: 10.1038/s41467-018-08284-6

[40] Moro F, Ke S, del Águila AG, Söll A, Sofer Z, Wu Q, et al. Revealing 2D magnetism in a bulk CrSBr single crystal by electron spin resonance. Advanced Functional Materials. 2022;**32**:2207044. DOI: 10.1002/adfm.202207044

[41] Scheie A, Ziebel M, Chica DG, Bae YJ, Wang X, Kolesnikov AI, et al. Spin waves and magnetic exchange hamiltonian in CrSBr. Advancement of Science. 2022;**9**:2202467. DOI: 10.1002/advs.202202467

[42] Jiang Z, Wang P, Xing J, Jiang X, Zhao J. Screening and design of novel 2D ferromagnetic materials with high curie temperature above room temperature. ACS Applied Materials & Interfaces. 2018;**10**:39032-39039. DOI: 10.1021/acsami.8b14037

[43] Lee K, Dismukes AH, Telford EJ, Wiscons RA, Wang J, Xu X, et

al. Magnetic order and symmetry in the 2D semiconductor CrSBr. Nano Letters. 2021;**21**:3511-3517. DOI: 10.1021/acs.nanolett.1c00219

[44] López-Paz SA, Guguchia Z, Pomjakushin VY, Witteveen C, Cervellino A, Luetkens H, et al. Dynamic magnetic crossover at the origin of the hidden-order in van der Waals antiferromagnet CrSBr. Nature Communications. 2022;**13**:4745. DOI: 10.1038/s41467-022-32290-4

[45] Telford EJ, Dismukes AH, Lee K, Cheng M, Wieteska A, Bartholomew AK, et al. Layered antiferromagnetism induces large negative magnetoresistance in the van der Waals semiconductor CrSBr. Advanced Materials. 2020;**32**:2003240. DOI: 10.1002/adma.202003240

[46] Susilo RA, Jang BG, Feng J, Du Q, Yan Z, Dong H, et al. Band gap crossover and insulator–metal transition in the compressed layered $CrPS_4$. npj Quantum Materials. 2020;**5**:58. DOI: 10.1038/s41535-020-00261-x

[47] Lee J, Ko TY, Kim JH, Bark H, Kang B, Jung S-G, et al. Structural and optical properties of single-and few-layer magnetic semiconductor $CrPS_4$. ACS Nano. 2017;**11**:10935-10944. DOI: 10.1021/acsnano.7b04679

[48] Peng Y, Ding S, Cheng M, Hu Q, Yang J, Wang F, et al. Magnetic structure and metamagnetic transitions in the van der Waals antiferromagnet $CrPS_4$. Advanced Materials. 2020;**32**:2001200. DOI: 10.1002/adma.202001200

[49] Yang J, Fang S, Peng Y, Liu S, Wu B, Quhe R, et al. Layer-dependent giant magnetoresistance in two-dimensional $CrPS_4$ magnetic tunnel junctions. Physical Review Applied. 2021;**16**:024011. DOI: 10.1103/PhysRevApplied.16.024011

[50] Harms NC, Smith KA, Haglund AV, Mandrus DG, Liu Z, Kim H-S, et al. Metal site substitution and role of the dimer on symmetry breaking in $FePS_3$ and $CrPS_4$ under pressure. ACS Applied Electronic Materials. 2022;**4**:3246-3255. DOI: 10.1021/acsaelm.2c00563

[51] Gu P, Tan Q, Wan Y, Li Z, Peng Y, Lai J, et al. Photoluminescent quantum interference in a van der Waals magnet preserved by symmetry breaking. ACS Nano. 2020;**14**:1003-1010. DOI: 10.1021/acsnano.9b08336

[52] Wang H, Fan F, Zhu S, Wu H. Doping enhanced ferromagnetism and induced half-metallicity in CrI_3 monolayer. EPL. 2016;**114**:47001. DOI: 10.1209/0295-5075/114/47001

[53] Kanamori J. Super exchange interaction and symmetry properties of electron orbitals. Journal of Physics and Chemistry of Solids. 1959;**10**:87-98. DOI: 10.1016/0022-3697(59)90061-7

[54] Hansen WN. Some magnetic properties of the chromium (III) halides at 4.2°K. Journal of Applied Physics. 2009;**30**:S304-S305. DOI: 10.1063/1.2185944

[55] Narath A. Low-temperature sublattice magnetization of antiferromagnetic $CrCl_3$. Physics Review. 1963;**131**:1929-1942. DOI: 10.1103/PhysRev.131.1929

[56] Narath A, Davis HL. Spin-wave analysis of the sublattice magnetization behavior of antiferromagnetic and ferromagnetic $CrCl_3$. Physics Review. 1965;**137**:A163-A178. DOI: 10.1103/PhysRev.137.A163

[57] Kuhlow B. Magnetic ordering in $CrCl_3$ at the phase transition. Physica Status Solidi A: Applications and

Materials Science. 1982;**72**:161-168. DOI: 10.1002/pssa.2210720116

[58] Tsubokawa I. On the magnetic properties of a $CrBr_3$ single crystal. Journal of the Physical Society of Japan. 1960;**15**:1664-1668. DOI: 10.1143/JPSJ.15.1664

[59] Le Flem G, Brec R, Ouvard G, Louisy A, Segransan P. Magnetic interactions in the layer compounds MPX_3 (M = Mn, Fe, Ni; X = S, Se). Journal of Physics and Chemistry of Solids. 1982;**43**:455-461. DOI: 10.1016/0022-3697(82)90156-1

[60] Huang B, McGuire MA, May AF, Xiao D, Jarillo-Herrero P, Xu X. Emergent phenomena and proximity effects in two-dimensional magnets and heterostructures. Nature Materials. 2020;**19**:1276-1289. DOI: 10.1038/s41563-020-0791-8

[61] Joy PA, Vasudevan S. Magnetism in the layered transition-metal thiophosphates MPS_3 (M = Mn, Fe, and Ni). Physical Review B. 1992;**46**:5425-5433. DOI: 10.1103/PhysRevB.46.5425

[62] Olsen T. Magnetic anisotropy and exchange interactions of two-dimensional $FePS_3$, $NiPS_3$ and $MnPS_3$ from first principles calculations. Journal of Physics D: Applied Physics. 2021;**54**:314001. DOI: 10.1088/1361-6463/ac000e

[63] Susner MA, Chyasnavichyus M, McGuire MA, Ganesh P, Maksymovych P. Metal thio- and selenophosphates as multifunctional van der Waals layered materials. Advanced Materials. 2017;**29**:1602852. DOI: 10.1002/adma.201602852

[64] Jernberg P, Bjarman S, Wäppling R. $FePS_3$: A first-order phase transition in a "2D" ising antiferromagnet. Journal of Magnetism and Magnetic Materials. 1984;**46**:178-190. DOI: 10.1016/0304-8853(84)90355-X

[65] Rule KC, McIntyre GJ, Kennedy SJ, Hicks TJ. Single-crystal and powder neutron diffraction experiments on $FePS_3$: Search for the magnetic structure. Physical Review B. 2007;**76**:134402. DOI: 10.1103/PhysRevB.76.134402

[66] Wildes AR, Simonet V, Ressouche E, McIntyre GJ, Avdeev M, Suard E, et al. Magnetic structure of the quasi-two-dimensional antiferromagnet $NiPS_3$. Physical Review B. 2015;**92**:224408. DOI: 10.1103/PhysRevB.92.224408

[67] Göser O, Paul W, Kahle HG. Magnetic properties of CrSBr. Journal of Magnetism and Magnetic Materials. 1990;**92**:129-136. DOI: 10.1016/0304-8853(90)90689-N

[68] Son J, Son S, Park P, Kim M, Tao Z, Oh J, et al. Air-stable and layer-dependent ferromagnetism in atomically thin van der Waals $CrPS_4$. ACS Nano. 2021;**15**:16904-16912. DOI: 10.1021/acsnano.1c07860

[69] Pei QL, Luo X, Lin GT, Song JY, Hu L, Zou YM, et al. Spin dynamics, electronic, and thermal transport properties of two-dimensional $CrPS_4$ single crystal. Journal of Applied Physics. 2016;**119**:043902. DOI: 10.1063/1.4940948

[70] Louisy A, Ouvrard G, Schleich DM, Brec R. Physical properties and lithium intercalates of $CrPS_4$. Solid State Communications. 1978;**28**:61-66. DOI: 10.1016/0038-1098(78)90328-9

[71] Seyler KL, Zhong D, Klein DR, Gao S, Zhang X, Huang B, et al. Ligand-field helical luminescence in a 2D ferromagnetic insulator. Nature Physics.

2018;**14**:277-281. DOI: 10.1038/s41567-017-0006-7

[72] Wu M, Li Z, Cao T, Louie SG. Physical origin of giant excitonic and magneto-optical responses in two-dimensional ferromagnetic insulators. Nature Communications. 2019;**10**:2371. DOI: 10.1038/s41467-019-10325-7

[73] Kim SY, Kim TY, Sandilands LJ, Sinn S, Lee M-C, Son J, et al. Charge-spin correlation in van der Waals antiferromagnet $NiPS_3$. Physical Review Letters. 2018;**120**:136402. DOI: 10.1103/PhysRevLett.120.136402

[74] Wang X, Cao J, Lu Z, Cohen A, Kitadai H, Li T, et al. Spin-induced linear polarization of photoluminescence in antiferromagnetic van der Waals crystals. Nature Materials. 2021;**20**:964-970. DOI: 10.1038/s41563-021-00968-7

[75] Du K-z, Wang X-z, Liu Y, Hu P, Utama MIB, Gan CK, et al. Weak van der Waals stacking, wide-range band gap, and Raman study on ultrathin layers of metal phosphorus trichalcogenides. ACS Nano. 2016;**10**:1738-1743. DOI: 10.1021/acsnano.5b05927

[76] Lane C, Zhu J-X. Thickness dependence of electronic structure and optical properties of a correlated van der Waals antiferromagnetic $NiPS_3$ thin film. Physical Review B. 2020;**102**:075124. DOI: 10.1103/PhysRevB.102.075124

[77] Kang S, Kim K, Kim BH, Kim J, Sim KI, Lee J-U, et al. Coherent many-body exciton in van der Waals antiferromagnet $NiPS_3$. Nature. 2020;**583**:785-789. DOI: 10.1038/s41586-020-2520-5

[78] Wilson NP, Lee K, Cenker J, Xie K, Dismukes AH, Telford EJ, et al. Interlayer electronic coupling on demand in a 2D magnetic semiconductor. Nature Materials. 2021;**20**:1657-1662. DOI: 10.1038/s41563-021-01070-8

[79] Kim S, Yoon S, Ahn H, Jin G, Kim H, Jo M-H, et al. Photoluminescence path bifurcations by spin flip in two-dimensional $CrPS_4$. ACS Nano. 2022;**16**:16385-16393. DOI: 10.1021/acsnano.2c05600

[80] Zhang Z, Shang J, Jiang C, Rasmita A, Gao W, Yu T. Direct photoluminescence probing of ferromagnetism in monolayer two-dimensional $CrBr_3$. Nano Letters. 2019;**19**:3138-3142. DOI: 10.1021/acs.nanolett.9b00553

[81] Belvin CA, Baldini E, Ozel IO, Mao D, Po HC, Allington CJ, et al. Exciton-driven antiferromagnetic metal in a correlated van der Waals insulator. Nature Communications. 2021;**12**:4837. DOI: 10.1038/s41467-021-25164-8

[82] Ho C-H, Hsu T-Y, Muhimmah LC. The band-edge excitons observed in few-layer $NiPS_3$. npj 2D Materials and Applications. 2021;**5**:8. DOI: 10.1038/s41699-020-00188-8

[83] McCreary A, Mai TT, Utermohlen FG, Simpson JR, Garrity KF, Feng X, et al. Distinct magneto-Raman signatures of spin-flip phase transitions in CrI_3. Nature Communications. 2020;**11**:3879. DOI: 10.1038/s41467-020-17320-3

[84] Zhang Y, Wu X, Lyu B, Wu M, Zhao S, Chen J, et al. Magnetic order-induced polarization anomaly of Raman scattering in 2D magnet CrI_3. Nano Letters. 2020;**20**:729-734. DOI: 10.1021/acs.nanolett.9b04634

[85] Liu S, Granados del Águila A, Bhowmick D, Gan CK, Thu Ha Do T,

Prosnikov MA, et al. Direct observation of magnon-phonon strong coupling in two-dimensional antiferromagnet at high magnetic fields. Physical Review Letters. 2021;**127**:097401. DOI: 10.1103/PhysRevLett.127.097401

[86] Yin T, Ulman KA, Liu S, Granados del Águila A, Huang Y, Zhang L, et al. Chiral phonons and giant magneto-optical effect in $CrBr_3$ 2D magnet. Advanced Materials. 2021;**33**:2101618. DOI: 10.1002/adma.202101618

[87] Huang B, Cenker J, Zhang X, Ray EL, Song T, Taniguchi T, et al. Tuning inelastic light scattering via symmetry control in the two-dimensional magnet CrI_3. Nature Nanotechnology. 2020;**15**:212-216. DOI: 10.1038/s41565-019-0598-4

[88] Cenker J, Huang B, Suri N, Thijssen P, Miller A, Song T, et al. Direct observation of two-dimensional magnons in atomically thin CrI_3. Nature Physics. 2021;**17**:20-25. DOI: 10.1038/s41567-020-0999-1

Chapter 3

Perspective Chapter: Environmental Impact of Modern Permanent Magnets

Belqees Hassan

Abstract

Modern permanent magnets, such as neodymium-iron-boron (NdFeB) and samarium-cobalt (SmCo), have revolutionized many industries due to their high magnetic strength and stability. However, the production and disposal of modern permanent magnets have significant environmental impacts that must be addressed. To minimize these impacts, it is important to reduce our reliance on rare earth elements by developing alternative materials or improving recycling technologies. Because when modern permanent magnets reach the end of their useful life, they are often disposed of in landfills or incinerated. This can lead to the leaching of toxic metals into the environment or the release of harmful gases into the atmosphere. So, proper disposal methods should be implemented to prevent contamination of soil, water sources, and air. This chapter will explore the potential environmental impacts of modern permanent magnets, including their production, use, and disposal. It will also provide recommendations for minimizing these impacts.

Keywords: magnetic strength, environmental impacts, rare earth elements, recycling technologies, harmful gases

1. Introduction

Permanent magnets are found in a wide range of products, from small electronic devices to the largest wind turbines [1]. As their name suggests, permanent magnets are made from materials that can be magnetized and remain magnetized for long periods. The most common permanent magnet materials are iron, nickel, and cobalt. Permanent magnets are an essential component of many modern technologies, and their impact on the environment is both direct and indirect. Direct impacts occur when magnets are manufactured, used, or disposed of. Indirect impacts occur because of the electricity generated by magnet-powered devices [2]. The manufacture of permanent magnets generally relies on mining, which can have a significant impact on the environment [3]. For example, iron mining releases large amounts of dust and other particulates into the air, which can lead to respiratory problems for nearby residents. In addition, the disposal of magnet materials can be difficult, as they are not biodegradable and often contain rare earth metals, which are difficult to recycle.

The use of permanent magnets also has environmental impacts. Magnets are often used in electric motors, which can be a significant source of air pollution [4].

2. Recycling and disposal options of modern permanent magnets

In recent years, researchers have been exploring new ways to make permanent magnets that are more environmentally friendly. These approaches include using alternative materials, such as iron and cobalt, and developing new manufacturing processes that reduce waste and energy consumption [5, 6]. By adopting these new methods, we can create sustainable permanent magnets that support our technological advances without damage. New ways to reduce environmental damage from permanent magnets include:

2.1 Protecting the environment with recycling rare earth elements

One way to reduce the environmental impact of permanent magnets is to recycle rare earth elements from old or discarded magnets. This process involves extracting and purifying the rare earth elements from the magnet, which can then be used to create new magnets. Electronic waste, also known as e-waste, may contain rare earth elements [7]. These elements are used in the production of electronic devices such as smartphones, laptops, and televisions. When these devices are disposed of improperly, the rare earth elements they contain can end up in landfills or incinerators. This not only wastes valuable resources but also poses a risk to the environment and human health. Recycling e-waste is one way to recover rare earth elements and reduce the environmental impact of their disposal. However, the process of extracting these elements from e-waste can be challenging and expensive due to their low concentration and complex composition. Therefore, it is important to design electronic devices with recycling in mind and develop more efficient methods for recovering rare earth elements from e-waste [8]. There are several methods of rare earth element (REE) recycling, including:

1. Solvent extraction: This method involves dissolving the REE-containing material in a solvent and then extracting the REEs using another solvent. This process is commonly used in the recycling of neodymium and dysprosium from magnets [9].

2. Ion exchange: In this method, the REEs are separated from other metals by exchanging ions with a resin. This process is commonly used in the recycling of yttrium and europium from fluorescent lamps [10].

3. Precipitation: This method involves adding a chemical to the REE-containing material to precipitate out the REEs. The precipitate is then collected and processed further to extract the individual REEs [11].

4. Pyrometallurgy: This method involves heating the REE-containing material to high temperatures to separate out the individual REEs. This process is commonly used in the recycling of cerium and lanthanum from catalytic converters [12].

5. Hydrometallurgy: In this method, the REE-containing material is dissolved in an acid or base solution, and then the individual REEs are separated using various chemical processes [13].

Material	Flux density (T)	Coercivity (kA/m)	Maximum energy product (kJ/m^3)	Curie temperature (°C)
NdFeB	1.1–1.4	750–2000	200–440	310–400
SmCo	0.8–1.1	600–2000	120–240	720–800
AlNiCo	0.6–1.4	40–140	10–80	700–860
Ferrite	0.2–0.5	100–300	10–40	450

Table 1.
Compares some properties of permanent magnetic materials.

6. Biometallurgy: This method involves using microorganisms to extract and recover REEs from waste materials such as electronic waste or mine tailings [14].

In general, these methods can help reduce our dependence on mining for new sources of rare earth elements and promote sustainable practices in the industry. Recycled rare earth elements can be used in a variety of applications, including:

1. Electronics: Rare earth elements are used in the production of electronic devices such as smartphones, laptops, and televisions. Recycling these elements can help reduce the environmental impact of electronic waste [15]. **Table 1** compares some properties of permanent magnetic materials used in electrical machines. Modern permanent magnets, such as neodymium-iron-boron (NdFeB) and samarium-cobalt (SmCo), have revolutionized many industries due to their high magnetic strength and stability. However, the production and disposal of these magnets have significant environmental impacts [16, 17].

2. Renewable energy: Rare earth elements are used in the production of wind turbines, solar panels, and electric vehicle batteries. Recycling these elements can help reduce the reliance on mining for new materials [18].

3. Medical equipment: Rare earth elements are used in medical imaging equipment such as MRI machines and X-ray machines. Recycling these elements can help reduce the cost of medical equipment and improve access to healthcare [19].

4. Defense industry: Rare earth elements are used in the production of defense equipment such as missiles, radar systems, and night vision goggles. Recycling these elements can help reduce the dependence on foreign sources for these critical materials [20].

5. Lighting: Rare earth elements are used in the production of energy-efficient lighting such as light emitting diode (LED) bulbs. Recycling these elements can help reduce energy consumption and greenhouse gas emissions [21].

6. Overall, recycling rare earth elements is an important step toward creating a more sustainable and environmentally friendly future.

2.1.1 The method used in the manufacture of rare earth magnets

The elements used to make magnets are not always iron. Instead, magnets are usually made of alloys, including some rare metals. The most common of these rare metals is

Rare metals	Formula	Industry description permanent magnets
Neodymium (Nd)	$Nd_2 Fe_{14}B$	Strong magnetic & low cost
Praseodymium (Pr)	$Pr_2Fe_{14}B$	Magnetic strength & durability
Dysprosium (Dy)	$DyFe_2$ & $(Nd,Dy)_2Fe_{14}B$	High thermal tolerance
Terbium (Tb)	$TbFe_2$ and TbDyFe	High magnetic strength
Samarium (Sm)	$SmCo_5$ and Sm_2Co_{17}	High magnetic energy
Europium (Eu)	$EuFe_2B_2$	Corrosion resistance
Gadolinium (Gd)	$GdCo_5$ and Gd_2Co_{17}	High magnetic energy
Holmium (Ho)	$HoCo_5$ and Ho_2Co_{17}	High magnetic synergy
Yttrium (Y)	$Y_2Co_{14}B$	High-performance magnets

Table 2.
List of rare earth elements used to make magnets arranged in order of importance.

neodymium. Neodymium magnets are the strongest of the rare earth magnets. Other rare metals used in magnet production include dysprosium and samarium. Cobalt is another metal used in the production of magnets, as shown in **Table 2** [22, 23].

Different methods are used to manufacture rare earth magnets, but one of the most common ones is called powder metallurgy. This method involves the following steps, as shown in **Figure 1**.

The raw materials used in the manufacture of rare earth magnets are neodymium, iron, and boron. These materials are firstly crushed into small particles and then mixed together in a certain proportion. The mixed raw materials are then melted in a vacuum furnace at high temperatures to form a molten alloy. The molten alloy is then cast into a mold to form a solid block of material. The solid mass of the material is then ground into a fine powder using a ball mill. Followed by compacting the fine powder into the desired shape using a hydraulic press. Next comes the sintering process, where the pressed powder is heated in a vacuum or in an inert atmosphere to a high temperature (about 1000°C), allowing some of the elements to melt and bond together. This process is called liquid phase sintering and gives the magnet its final strength and density. The sintered magnet is then ground, sliced, or cut into its final shape. This can be done with diamond tools or wire saws. Finally, the magnet is coated with a protective layer to prevent corrosion and improve its performance. This method allows for precise control over the composition and microstructure of the magnet, resulting in magnets with high magnetic properties and excellent performance characteristics [24].

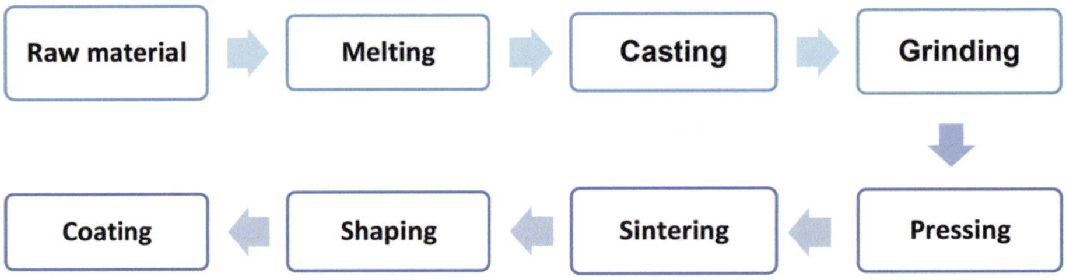

Figure 1.
The method used in the manufacture of rare earth magnets.

2.1.2 Challenges in recycling rare earth elements

There are many challenges associated with recycling rare earth elements. One major hurdle is that the amount of REE used in final products ranges from less than a milligram to several kilograms [25]. This means that assembly and disassembly of REE-containing devices can be difficult. In addition, earth-recycling methods require rare conventional hazardous chemicals such as hydrochloric acid, a lot of heat, and, therefore, a lot of energy [26].

2.2 Protecting the environment with biodegradable magnets

While magnetic materials have many beneficial applications, they also come with environmental concerns. Traditional magnetic materials, such as neodymium and samarium-cobalt, are not biodegradable and can have negative long-term effects on the environment if not disposed of properly. Biodegradable magnets have emerged as a promising alternative that can mitigate these environmental concerns [27, 28]. These materials are typically made from biodegradable polymers, such as polylactic acid or cellulose, mixed with magnetic particles such as iron oxide or ferrite. The advantages of biodegradable magnets include their ability to degrade naturally over time, reducing the amount of waste that ends up in landfills or oceans [29]. They also have lower toxicity levels and can be produced using more sustainable manufacturing processes, making them a more environmentally friendly option. In General, the use of biodegradable magnets has the potential to improve the sustainability of various industries that rely on magnetic materials, such as electronics, automotive, and medical industries.

2.2.1 Advantages and disadvantages of using biodegradable magnets

Biodegradable magnets have several advantages over traditional magnetic materials when it comes to protecting the environment.

1. It can degrade naturally over time without releasing any hazardous materials into the environment. This means that they can be safely disposed of without causing any long-term harm to ecosystems or human health.

2. Biodegradable magnets are generally made from renewable resources such as plant-based polymers or starch-based plastics that are less energy-intensive than traditional synthetic plastics or metals used in permanent magnet production processes [30].

3. Biodegradable magnets can be recycled more easily than traditional magnetic materials due to their ability to break down naturally over time without releasing any hazardous materials into the environment [31].

However, there are also some drawbacks associated with the use of degradable magnets that must be considered before determining whether they are suitable for a particular application. First, biodegradable magnets tend to have lower magnetic properties than conventional permanent magnetic materials, which may limit their usefulness in some applications that require high levels of magnetism (for example, medical imaging) [32]. Second, biodegradable magnets tend to be more expensive than traditional permanent magnet materials due to their higher production costs and shorter lives (i.e., they will need to be replaced more frequently). Finally, there is still a lot of research to be done before this material can be widely adopted by industry due to its relatively new nature and lack of commercial availability now (although this is likely to change in the future).

There has been considerable research conducted on biodegradable magnets in recent years with a focus on improving their magnetic properties while maintaining their environmental benefits (i.e., breaking down naturally without releasing any hazardous substances). For example, researchers have developed new types of polymers which can be used as binders for magnetic particles, which improve both strength and flexibility while still allowing them to break down naturally over time [33, 34]. Magnetic particles are a type of nanomaterial that can be used to improve the strength and flexibility of materials while still allowing them to break down naturally over time. These particles are typically composed of iron oxide, which is a magnetic material that can be manipulated by an external magnetic field. When these particles are added to a material, they create a network of tiny magnetic fields that interact with each other, increasing the strength and flexibility of the material [35, 36]. Additionally, these particles are biodegradable and will break down naturally over time, making them an environmentally friendly option for improving materials. Additionally, researchers have developed new types of starch-based plastics, which have been found to possess superior mechanical properties compared with traditional synthetic plastics while still being capable of breaking down naturally [37].

2.3 Reduce environmental damage with magnetic polymers

Magnetic polymers are a type of polymer that contains magnetic particles within their structure. These particles can be manipulated by an external magnetic field, making them useful in a variety of applications, including environmental remediation. One potential use for magnetic polymers is in reducing environmental damage caused by oil spills. When an oil spill occurs, it can have devastating effects on the surrounding ecosystem [38, 39]. The oil can coat the feathers of birds and fur of mammals, making it difficult for them to regulate their body temperature and leading to hypothermia. It can also contaminate water sources, killing fish, and other aquatic life. Magnetic polymers could be used to help clean up these spills by attracting and removing the oil from the water. The magnetic particles within the polymer would bind to the oil molecules, allowing them to be easily removed from the water using a magnet. This would reduce the amount of time and resources needed for cleanup efforts and minimize the impact on the environment. Another potential use for magnetic polymers is in removing heavy metals from contaminated soil or water. Heavy metals such as lead, mercury, and cadmium can be toxic to humans and wildlife, causing various health problems. Magnetic polymers could be used to selectively remove these metals from contaminated areas, reducing their impact on the environment [40]. Generally, the use of magnetic polymers has great potential for reducing environmental damage caused by various pollutants. As research continues in this area, we may see more widespread adoption of this technology in environmental remediation efforts [41]. This review article provides an overview of the use of magnetic polymers for environmental remediation, including their synthesis, properties, and applications in the removal of pollutants from water and soil [42]. This paper discusses the synthesis and characterization of magnetic polymer composites and their potential applications in environmental remediation, such as the removal of heavy metals from wastewater [43]. This research paper focuses on the synthesis and characterization of magnetic polymer nanocomposites for environmental applications, including their use in the removal of organic pollutants from water [44]. This study investigates the use of magnetic polymer microspheres for the removal of heavy metals from contaminated water, demonstrating their high efficiency in removing lead ions [45]. This research paper reports on the synthesis and characterization of magnetic polymer nanoparticles for environmental applications, including their use in the removal of organic pollutants from water and soil samples with high efficiency and selectivity. Overall, magnetic polymers represent a promising class of materials with unique properties that have potential applications in various fields.

2.4 Reduce environmental damage with magnetic nanoparticles

Magnetic nanoparticles have emerged as a promising tool for reducing environmental damage. These tiny particles, typically less than 100 nanometers in size, can be engineered to selectively bind to specific pollutants in water or soil [46–48]. Once bound, the nanoparticles can be easily removed using a magnetic field, leaving behind cleaner water or soil. One example of this technology in action is the use of magnetic nanoparticles to remove heavy metals from contaminated water. Heavy metals such as lead, mercury, and cadmium are toxic to humans and wildlife and can persist in the environment for decades. Magnetic nanoparticles functionalized with chelating agents can selectively bind to these metals and remove them from water. Another application of magnetic nanoparticles is in the remediation of oil spills. When oil spills occur, they can have devastating effects on marine ecosystems. Magnetic nanoparticles can be used to selectively bind to the oil droplets and then be removed using a magnetic field [49]. This approach is effective in laboratory studies and could potentially be used in real-world oil spill scenarios. Overall, the use of magnetic nanoparticles for environmental remediation shows great promise for reducing environmental damage caused by pollutants such as heavy metals and oil spills [50]. With continued research and development, the unique properties of magnetic nanoparticles make them a promising area of research for a wide range of applications [51].

2.5 Reduce environmental damage with renewable energy sources

Renewable energy sources can be replenished by nature and do not emit greenhouse gases or pollutants into the air [52]. Renewable energy sources include Solar Energy, where solar panels convert sunlight into electricity, which can be used to power homes and businesses. This reduces the need for fossil fuels and helps to reduce greenhouse gas emissions [53]. As well as wind Energy, Wind turbines generate electricity by harnessing the power of the wind. This is a clean and renewable energy source that does not produce harmful emissions [54]. In addition, renewable energy sources include hydro Energy; hydroelectric power plants generate electricity by using the force of moving water to turn turbines. This is a clean and renewable energy source that does not produce harmful emissions. Also Geothermal Energy, Geothermal power plants use heat from the earth's core to generate electricity. This is a clean and renewable source of energy that does not produce any harmful emissions [55]. Finally, Biomass Energy is another renewable source of energy; Biomass refers to organic matter such as wood, crops, or waste that can be burned to generate heat or electricity. This is a renewable source of energy that can help reduce greenhouse gas emissions [56]. Using these renewable energy sources can help reduce our dependence on fossil fuels and minimize environmental damage caused by burning coal, oil, and gas.

2.6 Reduce environmental damage with green manufacturing processes

Green manufacturing processes can play a significant role in reducing environmental damage caused by manufacturing activities. Green manufacturing refers to the use of environmentally sustainable practices in the production of goods. It involves reducing the use of raw materials, water, energy, and other resources and minimizing waste and greenhouse gas emissions [57]. Green manufacturing is a process that reduces environmental damage during the production of goods. This is because manufacturing processes often involve the use of harmful chemicals and materials, and the waste products created by manufacturing can be difficult to dispose of properly. Green manufacturing is a process that seeks to reduce the amount of environmental damage caused by manufacturing [58]. There are several ways in which green manufacturing can be achieved, as shown in **Figure 2**.

One way to reduce the environmental damage caused by manufacturing is to use alternative energy sources to power manufacturing processes, for example, solar, wind, and hydropower to power manufacturing facilities. This reduces the reliance on fossil fuels and helps to reduce greenhouse gas emissions [59]. Another way to reduce the environmental damage caused by manufacturing is to use recycled materials in the manufacturing process. This helps to conserve natural resources and reduces the amount of waste that ends up in landfills. In addition, green manufacturing processes involve the use of eco-friendly materials such as biodegradable plastics, recycled

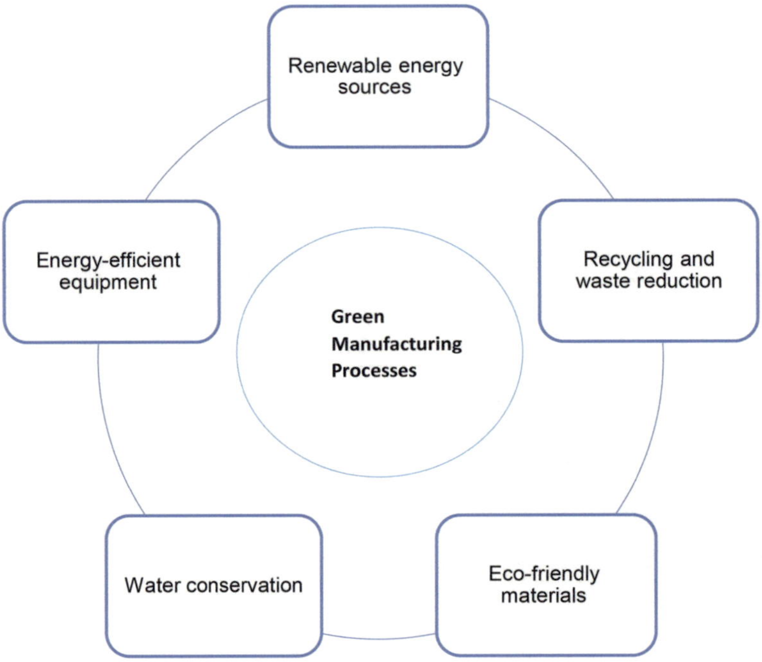

Figure 2.
Green manufacturing process that reduces environmental damage.

paper, and sustainable wood products. This reduces the environmental impact of manufacturing activities by reducing the use of nonrenewable resources [60]. Green manufacturing processes also involve water conservation techniques such as rainwater harvesting, water recycling, and wastewater treatment. This helps to conserve water resources and reduces water pollution [61]. Finally, green manufacturing processes involve the use of energy-efficient equipment such as Light Emitting Diode (LED) lighting, high-efficiency motors, and heating, ventilation, and air conditioning (HVAC) systems. This reduces energy consumption and helps to reduce greenhouse gas emissions [62]. In general, green manufacturing processes are essential for reducing environmental damage caused by industrial activities. By adopting these practices, manufacturers can help protect natural resources while also improving their bottom line through reduced costs associated with waste disposal and energy consumption.

2.7 Reduce environmental damage with magnet recycling programs

Magnet recycling programs are an effective way to reduce the environmental damage caused by the disposal of magnets. By implementing magnet recycling programs, these materials can be recovered and reused, reducing the need for new materials to be mined or manufactured. There are several ways that magnet-recycling programs can be implemented.

2.7.1 Magnets can be recycled into new magnets

The recycling of magnets is an excellent way to reduce environmental damage. By recycling magnets, we are able to reuse them in a variety of ways. This not only reduces the amount of waste that ends up in landfills, but it also helps to conserve resources. When magnets are recycled, where old or discarded magnets are broken down into their base materials and then reformed into new magnet products [63] first, they are crushed into smaller pieces. These smaller pieces are then sorted by their magnetic properties. The magnet that is most magnetic is known as the ferrite, while the least magnetic is known as the neodymium. Once the magnets are sorted, they are then cleaned and smelted down. This process helps to reduce waste while conserving resources and energy. A sustainable solution benefits both the environment and the economy [64].

2.7.2 Recycling magnets helps reduce environmental pollution

Magnet recycling is one of the most efficient ways to reduce environmental pollution. Magnets, like most materials, are made up of natural resources. By recycling magnets, we can help reduce pollution and conserve these resources. The process of magnet recycling involves several steps:

1. The magnets are collected from various sources such as electronic waste, motors, generators, and hard drives.

2. The collected magnets are sorted according to their type, size, and composition. After sorting, the magnets are crushed into small pieces and separated from other materials using magnetic separation techniques.

3. The separated magnet pieces are then cleaned and processed to remove any impurities or contaminants.

4. Finally, the purified magnet pieces are melted down and reformed into new magnet products or sold to manufacturers for use in other applications.

The process of recycling magnets is not only efficient in reducing environmental pollution, but it is also cost-effective. By recycling magnets, we can save money on raw materials and on energy costs. In addition, recycling magnets help to create jobs in the recycling industry [65, 66].

3. Environmental regulations of modern permanent magnets

Modern permanent magnets, such as neodymium, samarium-cobalt, and ferrite magnets, are subject to environmental regulations. These regulations aim to minimize the environmental impact of the production, use, and disposal of these magnets.

3.1 The importance of environmental regulations in the production of modern permanent magnets

Typically, the environmental performance of high-performance permanent magnets is better than that of traditional magnets. This is due to the fact that high-performance permanent magnets are more efficient, which results in less energy consumption and lower greenhouse gas emissions during their use [67]. Additionally, high-performance magnets are often made using advanced manufacturing techniques that reduce waste and improve resource efficiency. However, it is important to note that high-performance magnets often contain rare earth elements, which can have a significant environmental impact during their mining and processing. Rare earth elements are often found in low concentrations, which means that a large amount of ore needs to be mined and processed to extract them. This can result in significant environmental damage, including soil erosion, water pollution, and habitat destruction. To mitigate these environmental impacts, efforts are being made to develop more sustainable methods for mining and processing rare earth elements. For example, some companies are exploring the use of more environmentally friendly solvents and reducing the amount of waste generated during processing [68]. Additionally, efforts are being made to recycle rare earth elements [69, 70] from end-of-life products, which can reduce the need for new mining and processing activities. Generally, while high-performance permanent magnets offer better environmental performance than traditional magnets during use, there is still work to be done to address the environmental impacts associated with their production.

3.2 Some environmental regulations of permanent magnets

Permanent magnets are present in many electronic devices, from the speakers in your cell phone to the motors in your car. The Materials Research Society notes, permanent magnets are fundamental to technologies as diverse as electric motors, generators, hard disk drives, medical devices, and magnetic resonance imaging. Permanent magnets are made from a variety of materials, including iron, nickel, cobalt, and rare earth metals. The manufacturing process of permanent magnets can release harmful chemicals and pollutants into the environment. In order to protect the environment and human health, various environmental regulations have been put in place regarding the production of permanent magnets. Some of the key environmental regulations of permanent magnets include the following:

1. Restriction of Hazardous Substances Regulations (RoHS): Permanent magnets may contain hazardous materials, such as lead, mercury, and cadmium, in electronic and electrical equipment that require specific handling and disposal procedures to comply with environmental regulations [71].

2. Registration, Evaluation, Authorization and Restriction of Chemicals (REACH) Regulation: This regulation aims to protect human health and the environment from the risks posed by chemicals. It requires companies to register the chemicals they produce or import and provide information on their properties and uses [72].

3. End-of-life disposal regulations: When permanent magnets reach the end of their useful life, they must be disposed of properly to minimize their impact on the environment. This may involve recycling, repurposing, or proper disposal in a designated landfill [73].

4. Manufacturing regulations: The production process for permanent magnets may involve the use of hazardous materials or emissions that need to be monitored and controlled to comply with environmental regulations [74].

5. Energy efficiency regulations: Permanent magnets are often used in motors and other energy-consuming devices and, therefore, may be subject to energy efficiency regulations that promote the use of more efficient devices to reduce energy consumption and emissions [75].

Both the RoHS and the REACH are important for the environmental protection of permanent magnets. These regulations help to ensure that permanent magnets are produced without the use of hazardous substances. As the world progresses, society is developing new ways to be more environmentally conscious. One way is by modernizing the environmental regulations of permanent magnets. New regulations require manufacturers to use less electricity, create fewer harmful emissions, and recycle magnets more often. This is beneficial for both the environment and the economy. Environmentally friendly practices are not only good for the environment, but they also save money in the end. Overall, it is important to consider the environmental impact of permanent magnets throughout their entire lifecycle, from production to disposal, and to comply with all applicable regulations to minimize their impact on the environment.

4. Recommendations for minimizing environmental impact of modern permanent magnets

1. Reduce the use of rare earth metals: Rare earth metals are essential components in modern permanent magnets, but their extraction and processing have significant environmental impacts. Therefore, reducing the use of rare earth metals in magnet production can help minimize the environmental impact.

2. Increase recycling: Recycling permanent magnets can help reduce the demand for new magnet production and minimize waste generation. Recycling also reduces the need for mining and processing raw materials, which can have significant environmental impacts.

3. Use renewable energy sources: The production of permanent magnets requires a significant amount of energy, which often comes from nonrenewable sources such as coal and oil. Using renewable energy sources such as solar or wind power can help reduce greenhouse gas emissions and minimize the environmental impact.

4. Improve manufacturing processes: Modern manufacturing processes can be optimized to reduce waste generation, water usage, and energy consumption. By improving manufacturing processes, it is possible to minimize the environmental impact of permanent magnet production.

5. Develop alternative materials: Researchers are exploring alternative materials that could replace rare earth metals in permanent magnets. Using alternative materials for magnet production, such as biomaterials, can help reduce the environmental impact of modern permanent magnets and could have lower environmental impacts than rare earth metals, which are more sustainable in the end.

6. Promote responsible sourcing: Companies that produce permanent magnets should ensure that they source their raw materials responsibly and ethically. This includes ensuring that mining practices are environmentally sustainable and that workers are treated fairly throughout the supply chain.

7. Efficient use of magnets: The efficient use of magnets can help reduce the amount of energy required for various applications, ultimately reducing greenhouse gas emissions.

8. End-of-life management: Proper end-of-life management of magnets, such as recycling or disposal in an environmentally responsible manner, is crucial for reducing their environmental impact.

9. Monitor compliance: Governments should monitor compliance with environmental regulations through regular inspections, audits, and reporting requirements.

10. Educate consumers: Consumers should be educated about the environmental impact of permanent magnets and encouraged to choose products that are environmentally friendly.

5. Conclusion

In conclusion, the environmental impact of modern permanent magnets is a complex issue that requires careful consideration. While these magnets are essential components in many modern technologies, their production and disposal can have negative impacts on the environment. However, there are several strategies that can be employed to minimize this impact, such as recycling, using alternative materials, adopting sustainable mining practices, efficient use of magnets, and proper end-of-life management. By implementing these strategies, we can minimize the environmental impact of modern permanent magnets. It is important for individuals and industries to be aware of the environmental impact of modern permanent magnets and take steps toward reducing their negative effects and move toward a more sustainable future.

Conflict of interest

"The authors declare no conflict of interest."

Author details

Belqees Hassan[1,2]*

1 Department of Physics, College of Science and Arts, Qassim University, Alasyah, Saudi Arabia

2 Physics Department, Faculty of Science, Ibb University, Ibb, Yemen

*Address all correspondence to: belqees.ahmad@yahoo.com

References

[1] Gutfleisch O, Willard MA, Brück E, Chen CH, Sankar SG, Liu JP. Magnetic materials and devices for the 21st century: Stronger, lighter, and more energy efficient. Advanced Materials. 2011;**23**(7):821-842. DOI: 10.1002/adma.201002180

[2] Shackelford JF. Introduction to Materials Science for Engineers. London, UK: Pearson; 2017

[3] Duckworth RC. Environmental impacts of industrialization in magnets. In: Handbook of Magnetic Materials. Vol. 26. Amsterdam, Netherlands: Elsevier; 2017. pp. 117-159

[4] Sharma A, Yadav S. Environmental impacts of rare earth mining and separation based on Eudialyte: A new European approach. Journal of Cleaner Production. 2019;**218**:1114-1129

[5] Jia H, Gao F. Advances in environmentally friendly permanent magnets. Rare Metals. 2019;**38**(10):849-858. DOI: 10.1007/s12598-019-01201-1

[6] Zhang Y, Zhao Y, Ma B, Yan M. Review on the research progress of the eco-friendly permanent magnet materials. Journal of Magnetism and Magnetic Materials. 2021;**527**:167596. DOI: 10.1016/j.jmmm.2020.167596

[7] Li J, Zhang Y, Wang X. Recycling of rare earth elements: A review of current technologies and future prospects. Inorganics. 2018;**6**(4):127. DOI: 10.3390/inorganics6040127

[8] Mishra S, Kapsalas S, Gervais D. Recycling rare earth elements from permanent magnets: A critical review. Journal of Sustainable Metallurgy. 2019;**5**(4):646-667. DOI: 10.1007/s40831-019-00239-6

[9] Rahman MA, Souders JE, Misra S, McElroy AE. Recycling rare earth elements from end-of-life products: A comprehensive review. Journal of Sustainable Metallurgy. 2020;**6**(4):721-750. DOI: 10.1007/s40831-020-00289-3

[10] Tunsu C, Ekberg C, Retegan T. Recovery of rare earth elements from fluorescent lamps: A review. Journal of Cleaner Production. 2016;**127**:362-375. DOI: 10.1016/j.jclepro.2016.03.032

[11] Jordens A, Cheng YP, Waters KE. A review of the beneficiation of rare earth element bearing minerals. Minerals Engineering. 2013;**41**:97-114. DOI: 10.1016/j.mineng.2012.10.014

[12] Liu H, Liao W. Extraction of rare earth elements from waste cathode ray tubes by pyrometallurgical processing. Journal of Hazardous Materials. 2017;**323**:563-570. DOI: 10.1016/j.jhazmat.2016.11.057

[13] Molycorp. Rare Earth Elements: Hydrometallurgy. 2012. Available from: https://www.molycorp.com/rare-earths/hydrometallurgy/

[14] Tran TV, Tran YT, Pham-Huu C. Biometallurgy: A sustainable method for rare earth extraction? Sustainability. 2019;**11**(4):1114. DOI: 10.3390/su11041114

[15] Jha A, Ranganathan S, Misra S. Rare earth elements: A review of applications, occurrence, exploration, analysis, recycling, and environmental impact. Resources, Conservation and Recycling. 2016;**115**:201-214

[16] Croat J, Ormerod J. Modern Permanent Magnets. 1st ed. Elsevier; 2022. DOI: 10.1016/C2020-0-03162-2

[17] Li W, Zhang X, Li J. Magnetic properties and applications of rare earth permanent magnetic materials: A review. Journal of Rare Earths. 2017;**35**(6):534-545. DOI: 10.1016/j.jre.2017.02.007

[18] Mishra S, Mishra T, Mishra BK, Lee J, Kim J. Recycling of rare earth elements: A sustainable solution for future energy technologies. Environmental Science & Technology. 2021;**55**(4):2137-2154

[19] Hurst KE. Rare earth elements in health care: A review of their use in diagnosis, treatment, and drug delivery. Journal of Rare Earths. 2019;**37**(2):123-134. DOI: 10.1016/j.jre.2018.09.014

[20] Feickert A. Rare Earth Elements in National Defense: Background, Oversight Issues, and Options for Congress. Congressional Research Service. Washington, D.C.: United States Government Publishing Office; 2019

[21] Erdmann R, Graedel TE. Criticality of non-fuel minerals: A review of major approaches and analyses. Environmental Science & Technology. 2011;**46**(17):8849-8858

[22] Kumar R, Murty BS. Magnetic materials: A review of the state of the art. Journal of Materials Science. 2006;**41**(15):4369-4383

[23] Wei Y. Rare Earth-Based Corrosion Inhibitors. Amsterdam, Netherlands: Elsevier; 2014

[24] Buschow KHJ, Magnus F. Handbook of Magnetic Materials. Vol. 27. Amsterdam: Elsevier; 2017. p. 3

[25] Habib MA, Sun W. Recycling rare earth elements: A need for sustainable approach. Resources, Conservation and Recycling. 2019;**144**:293-302. DOI: 10.1016/j.resconrec.2019.01.002

[26] Fujita Y, McCall SK, Ginosar D. Recycling rare earths: Perspectives and recent advances. MRS Bulletin. 2022;**47**:283-288. DOI: 10.1557/s43577-022-00301-w

[27] Binnemans K, Jones P, Blanpain B, Van Gerven T, Yang Y, Walton A, et al. Environmental impacts of rare earth mining and separation based on eudialyte: A new European way. Journal of Cleaner Production. 2015;**99**:290-300

[28] Sprengel S, Küfer KH, Andersen O, Schaub J. Environmental concerns of permanent magnet motors and alternatives to rare earth magnets. Proceedings of the IEEE. 2014;**102**(4):604-618

[29] Nlebedim IC, Maity T. Biodegradable magnets. Materials Today Chemistry. 2018;**8**:72-83. DOI: 10.1016/j.mtchem.2018.09.002

[30] Huang H et al. Biodegradable magnets: A comprehensive review. Journal of Materials Chemistry A. 2021;**9**(26):13752-13766. DOI: 10.1039/D1TA03284E

[31] Pal J et al. Biodegradable magnets: A green and sustainable alternative. ACS Sustainable Chemistry & Engineering. 2020;**8**(2):1007-1021. DOI: 10.1021/acssuschemeng.9b06176

[32] Sridhar S, Ramakrishna BR. Biodegradable magnetic nanoparticles and their applications in medicine. Journal of Nanomedicine & Nanotechnology. 2015;**6**(4):1-9. DOI: 10.4172/2157-7439.1000310

[33] Kumar A, Sharma S, Sharma P. Synthesis and characterization of novel polymers as binders for magnetic particles: Improving strength and flexibility. Polymer International. 2017;**66**(8):945-952

[34] Wang H, Li X, Wang S. Biodegradable magnetite nanocomposites for targeted drug delivery and magnetic resonance imaging. ACS Applied Materials & Interfaces. 2016;**8**(37):24607-24614

[35] Zhang Y, Li Y, Wang S. Biodegradable magnetite nanoparticles for magnetic resonance imaging and drug delivery. Journal of Materials Chemistry B. 2018;**6**(30):5134-5143

[36] Kimura T, Matsumoto K, Uchida M. Biodegradable magnetite nanoparticles as a contrast agent for magnetic resonance imaging: Preparation and in vivo evaluation in mice with a tumor model system using an intraperitoneal injection route. International Journal of Nanomedicine. 2015;**10**(1):3333-3342

[37] Liu J, Chen W, Yang L, Liu J, Zhang X, Sunagawa M. Magnetic biodegradable polymeric nanocomposites for targeted drug delivery and magnetic resonance imaging: A review of recent progress and perspectives on their applications in cancer therapy and diagnosis. International Journal of Nanomedicine. 2015;**10**(1):3333-3342

[38] Gao Y, Wang X, Li J, Zhang Y. Starch-based plastics: A review of properties and applications. Polymers. 2018;**10**(3):243

[39] Naik RR, Ramamurthy PC. Magnetic polymer composites for environmental applications. Chemical Engineering Journal. 2013;**225**:1-22. DOI: 10.1016/j.cej.2013.03.035

[40] Chen S. Magnetic polymer adsorbents for environmental remediation. Journal of Hazardous Materials. 2020;**384**:121390. DOI: 10.1016/j.jhazmat.2019.121390

[41] Shafiullah BM, Halim MA. Magnetic polymers for the removal of heavy metals from wastewater. Journal of Environmental Chemical Engineering. 2017;**5**(1):1132-1142. DOI: 10.1016/j.jece.2017.01.024

[42] Kadam SS, Kulkarni SB. Magnetic polymers for environmental remediation: A review. Polymer-Plastics Technology and Engineering. 2017;**56**(8):891-909. DOI: 10.1080/03602559.2016.1261870

[43] Uddin MA, Islam MR. Magnetic polymer composites for environmental applications. In: Encyclopedia of Renewable and Sustainable Materials. Elsevier; 2016. pp. 1-10

[44] Kaur A, Kumar R. Magnetic polymer nanocomposites for environmental applications. In: Handbook of Nanomaterials for Industrial Applications. Elsevier; 2019. pp. 827-854

[45] Lin Y-H, Wang H-P, Lai W-C, Chen Y-W. Magnetic polymer microspheres for environmental remediation. Journal of Hazardous Materials. 2016;**304**:527-539

[46] Kim S-H, Park J-S, Kim H-S, Lee J-H. Magnetic polymer nanoparticles for environmental applications: Synthesis, characterization, and performance evaluation. Journal of Nanomaterials. 2015;**2015**(573891):13

[47] Singh SK, Srivastava SK. Magnetic nanoparticles for environmental remediation: A review. Journal of Water Process Engineering. 2018;**24**:1-14

[48] Rahman MM, Asmatulu R, Fakhri A. Magnetic nanoparticles for environmental applications: A review. Journal of Nanoscience and Nanotechnology. 2019;**19**(3):1263-1282

[49] Lee J-H, Kim Y-J, Cho M. Magnetic nanoparticles in environmental analysis:

A review. Journal of the Korean Chemical Society. 2018;**62**(4):259-268

[50] Singh SK, Srivastava V, Singh P, Singh RP, Singh RP. Magnetic nanoparticles for oil spill cleanup: A review. Journal of Dispersion Science and Technology. 2019;**40**(6): 827-840

[51] Kim H-J, Kim H-J, Lee G, Kim K-J. Magnetic nanoparticles for environmental monitoring: A review. Journal of Industrial and Engineering Chemistry. 2017;**55**:1-13

[52] Singh SK, Srivastava SK. Magnetic nanoparticles for environmental remediation: A review. Journal of Water Process Engineering. 2018;**22**:249-258

[53] Hossain MA, Paul SK, Islam MR. Renewable energy sources and environmental sustainability: A review of the literature. Journal of Environmental Science and Natural Resources. 2015;**8**(1):33-41

[54] Kumaravelu V, Sundararajan R. Renewable energy sources for environmental protection: An overview. International Journal of Renewable Energy Research (IJRER). 2016;**6**(1):1-11

[55] Ahmed J, Wen P, Zhang X. Environmental benefits of renewable energy sources: A review. Renewable and Sustainable Energy Reviews. 2019;**107**:163-190

[56] Kumar S, Kaushik ND, Sharma S. Renewable energy sources for environmental protection: A case study of India. Journal of Cleaner Production. 2019;**209**:1112-1125

[57] Kothari S, Sharma SC, Tyagi SK. Renewable energy sources for environmental protection: A review. Renewable and Sustainable Energy Reviews. 2009;**13**(8):2045-2054

[58] Islam SM, Eltawil MA, Hassan MF. Green manufacturing: A review of sustainable manufacturing processes and practices. International Journal of Sustainable Energy. 2013;**32**(1-2): 47-56

[59] Sarkis J. Sustainable manufacturing: challenges and opportunities. International Journal of Production Economics. 2012;**140**(1):1-7

[60] Lee Y-C, Genovese A, Lenny Koh SC. Sustainable manufacturing: A review of recent developments. Journal of Cleaner Production. 2015;**97**:60-74

[61] Holmberg J, Robèrt B. Sustainable manufacturing: A critical review. Journal of Cleaner Production. 2000;**8**(6):504-517

[62] Chung S-H, Morris B, Ross J. Green Manufacturing: Principles and Practices. CRC Press; 2013

[63] Kulkarni R, Anantharaman S. Green manufacturing: An overview. International Journal of Advanced Manufacturing Technology. 2015;**76**(9-12):1673-1684

[64] Ramana KV, Groot MT, Binnemans K, Sahu R. A review on the recycling of rare earth metals from magnetic waste streams. Resources, Conservation and Recycling. 2018;**129**:53-68. DOI: 10.1016/j.resconrec.2017.10.039

[65] Xu G, Zhang L, Tang C, Chen K, Sun S. Recycling of rare earth magnets from hard disk drives: A review. Journal of Cleaner Production. 2019;**228**:1488-1500. DOI: 10.1016/j.jclepro.2019.04.317

[66] Binnemans K, Jones PT, Blanpain B, Van Gerven T, Yang Y, Walton A, et al. Recycling of permanent magnets: A critical review. Journal of Sustainable Metallurgy. 2013;**1**(1):27-42

[67] Ma G, Cai W, Zhang J. Environmental regulations and their impact on the rare earth industry. Journal of Cleaner Production. 2018;**190**:211-218. DOI: 10.1016/j.jclepro

[68] Zhang Y, Xu M, Keoleian GA. Greening the supply chain: A life cycle assessment of magnets. Environmental Science & Technology. 2014;**48**(21):12621-12628. DOI: 10.1021/es502063g

[69] Hwang E, Owens JD. Environmental regulations and the global supply chain for rare earth magnets. Resources, Conservation and Recycling. 2019;**146**:494-503. DOI: 10.1016/j.resconrec.2019.03.019

[70] Li J, Zhang L, Liu Y. Environmental regulation and sustainability in the production of rare-earth permanent magnets. Journal of Cleaner Production. 2020;**241**:118174. DOI: 10.1016/j.jclepro.2019.118174

[71] Walker LR, Kummer JG. Assessment of environmental impacts of permanent magnet motor systems. Journal of Cleaner Production. 2018;**172**:2092-2102

[72] Xu ZY, Zhang L, Cheng YT. Environmental impact analysis of rare earth permanent magnets production in China. Journal of Rare Earths. 2013;**31**(8):787-793

[73] Chen D-F, Habibullah MS, Yang J-M, Yang M-H. Environmental life cycle assessment of rare-earth permanent magnet generators for wind turbines. Renewable and Sustainable Energy Reviews. 2014;**34**:409-416

[74] Støckel K, Steenari BM, Akbari A. Environmental performance of high-performance permanent magnets. Journal of Cleaner Production. 2019;**208**:1120-1129

[75] Li J, Long X, Chen H, Zhang Y. Environmental regulation and green innovation in China's rare earth industry. Journal of Cleaner Production. 2015;**108**:1378-1387

Chapter 4

Preisach Hysteresis Model – Some Applications in Electrical Engineering

Alfredo Bermúdez, Dolores Gómez and Pablo Venegas

Abstract

In this chapter we recall the well-known hysteresis Preisach model, widely employed in the area of magnetism. Some applications of this model in electrical engineering are also described, with a specific focus on the estimation of electromagnetic losses in electrical machines, the simulation of magnetization-demagnetization processes arising in magnetic particle inspection, and the mathematical modeling of batteries for electric vehicles.

Keywords: magnetic hysteresis, Preisach model, electromagnetic losses, magnetic particle inspection, batteries for electric vehicles

1. Introduction

Hysteresis is a very complex nonlinear behavior affecting many physical phenomena. Systems affected by this behavior are characterized by the fact that the way they evolve in response to a stimulus depends not only on the cause of the stimulus, but also on the preceding states of the system. Thus, the same instantaneous values of the input can give different outputs depending on the history of the input applied, which gives rise to a relationship that is not only nonlinear but also multivalued, making it very difficult to model and control. This memory-based property is found in various areas of science and engineering such as mechanics, biology, economics and multiphase flow in porous media or magnetism, among others. It was precisely in the latter that the term was initially coined since most ferromagnetic materials exhibit hysteresis [1]. This means that when the material is subjected to the application of a magnetic field, the magnetic induction reached at each point of the material depends not only on the intensity of the applied field at a given instant of time, but also on its previous magnetic history. Despite the difficulties involved in its study, having a good hysteresis model is essential in a multitude of applications in the field of electrical engineering; an example of this is the estimation of energy losses in electrical machines.

The variety of systems with hysteresis is very broad and there is a large amount of bibliography devoted to hysteresis models in different communities of physicists, engineers, and mathematicians, where great efforts have been devoted to their development and analysis. As a result, nowadays there are several hysteresis models, ranging from simple to complex, each of them valid in some specific situation; see, e.g., the monographs and reviews [2–5].

The mathematical approach tries to deal with this phenomenon under a common mathematical framework, but this is not always possible. We highlight the monograph by Krasnosel'ski˘i and Pokrovski˘i [2], who have conceptually introduced the notion of hysteresis operator and carried out a systematic analysis of its properties. More recently, research on hysteresis models as well as their coupling with partial differential equations has been progressing; see, among others, the works by Visintin [5, 6] and Brokate and Sprekels [7]; from the physical point of view, Mayergoyz [8] and Bertotti [9] are the classical references.

Hysteresis models can be classified into two main classes according to whether or not the system response depends on the input velocity. In *static* or *rate-independent* models, output values are not affected by the velocity of the input but only by the values in the input range and by the order in which these values have been attained. Consequently, the model cannot reflect the frequency or waveform dependence of the field. This is a characteristic of the classical hysteresis phenomenon and in this sense some authors [6] consider hysteresis as a rate-independent memory only. On the opposite side are the so-called *dynamic* or *rate-dependent* models, which account for the effect of the velocity of changes in the applied input.

The so-called classical Preisach operator [10] is a rate-independent model originally designed to model hysteresis of ferromagnetic materials which is based on physical assumptions derived from the concept of magnetic domains [8, 9, 11, 12]. It is suitable for modeling scalar hysteresis and for this purpose, the magnetic field strength H is used as the input variable while the magnetization M (or the magnetic flux density strength, B) acts as the output variable. Its main advantage is the ability to describe not only the major hysteresis loop, but also inner loops and other complex characteristics of the magnetization processes. There are several extensions of this model that consider the rate of change of the stimulus to be able to consider phenomena where this factor influences the magnetization. These are generically referred to as *dynamic Preisach models* (see, for instance [13–17]).

Currently, the Preisach-type operators [18] are used not only in the area of magnetism, but also to describe hysteresis phenomena in other fields as diverse as fluid flow in porous media [19], elastoplasticity [20], solid phase transitions [7], shape memory alloys [21], or biology [22], to name a few.

The mathematical and numerical treatment of PDEs with hysteresis operators is a challenging issue because, despite the importance of the topic, still few results are available. Due to the strong interdisciplinary character of hysteresis phenomena, the interest for their study is continuously increasing in a great variety of applications. During the last few years, the authors of this chapter have been working on the mathematical modeling, numerical analysis and computation of PDE hysteresis-related problems motivated by real applications, with special emphasis on the hysteresis of ferromagnetic materials. This work is a survey of previous co-authored studies [17, 23–25], where we have intended to provide original steps into the mathematical and numerical treatment of parabolic problems with hysteresis. In all of them, the Preisach model was considered as hysteresis operator. The problems were addressed in different aspects: mathematical analysis, numerical methods, until convergence of approximate solutions and computational results to confirm the theory and to compare with experimental data. The rigorous mathematical framework of the numerical analysis are not included here, and we will be concerned with the results obtained for different industrial problems.

The chapter is organized as follows. In Section 2, we start by recalling the basic properties of the hysteresis operators. The well-known Preisach hysteresis model, both in the rate-independent and rate-dependent cases, is introduced in Section 3. Finally, in Section 4 we show some examples of the applications of the Preisach model in electrical engineering. These examples include the computation of hysteresis losses, the numerical simulation of magnetization and demagnetization processes in ferromagnetic workpieces and the simulation of batteries for electric vehicles.

2. Hysteresis operators: Basic properties

In this section, we summarize the basic knowledge of scalar hysteresis operators based on the description given in [6].

Most systems experiencing hysteresis phenomena display *hysteresis loops*. To illustrate this in a simple setting, let us consider a system like the one depicted in **Figure 1** (left), whose state is characterized by two time-dependent scalar variables, u and w. Let us also suppose that u is the independent variable, and thus the evolution of w depends on u. Such diagrams implicitly show the *memory effect* inherent to the hysteresis: at any time $t > 0$, $w(t)$ depends on the evolution of u in $[0, t]$, rather than only on $u(t)$. In many cases, not all the information in $[0, t]$ is used to compute the value of w but only the one which has not been *wiped-out* from the memory (see Section 3.1). In other words, the state of the system at present time depends on the previous *history* of its state. Consequently, the same instantaneous values of the input can give different outputs and when describing the relation between w and u in the (u, w) plane a multi-branch nonlinearity appears with branch to branch transitions (see **Figure 1**, left).

According to Ref. [6], two main characteristics of hysteresis phenomena can be distinguished: *causality* and *rate independence*. To clarify these concepts, we consider the (u, w) relation introduced above. *Causality* means that, at any time t, the value of $w(t)$ only depends on the previous evolution of u, namely, $w(t)$ depends on $u|_{[0,t]}$. On the other hand, *rate independence* is translated by the condition that, at any instant t, $w(t)$ depends just on the range of function $u : [0, t] \to \mathbb{R}$ and on the order in which the values of u have been attained. In other words, w is independent of the *velocity* of u. An example is given in **Figure 1** (right) where it is shown that three different functions $u : [0, t] \to \mathbb{R}$ lead to the same $w - u$ curve.

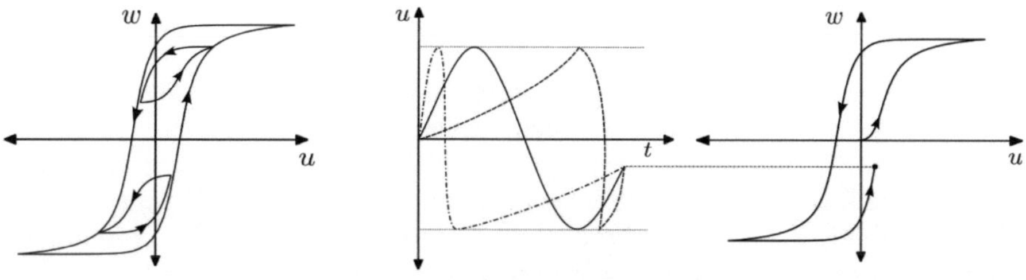

Figure 1.
Hysteresis loop (left) and rate independence example (right).

We notice that, even in the most typical hysteresis phenomena, like plasticity, ferroelectricity or ferromagnetism, the memory effect is not completely rate-independent, since viscous-type effects are coupled with hysteresis. However, in the applications presented in the following sections, this effect is neglected, so that the rate-independent component prevails.

To provide a functional setting for hysteresis operators, we first notice that, at any instant t, $w(t)$ will depend not only on the previous evolution of u (i.e., on $u|_{[0,t]}$), but also on the "initial state" of the system. Due to the memory dependence of hysteresis processes, additional information is needed to make up for the lack of history when the process begins. This initial information must represent the "history" of the function u before $t = 0$. Hence, not only the standard initial value $(u(0), w(0))$ must be provided. In general, a variable ξ containing all the information about the "initial state" is considered. This can be expressed, for instance, as follows:

$$(u, \xi) \rightarrow w = \tilde{\mathcal{F}}(u, \xi). \tag{1}$$

In the case of partial differential equations, it is necessary to define an operator $\mathcal{F}$ acting between suitable function spaces involving the space variable. Let $\hat{\Omega}$ be an open subset of $\mathbb{R}^N (N \geq 1)$; given a hysteresis operator $\tilde{\mathcal{F}}$, we introduce, for any u: $\hat{\Omega} \times [0, T] \rightarrow \mathbb{R}$ and any $\xi : \hat{\Omega} \rightarrow Y$, with Y a suitable space, the corresponding space dependent operator $\mathcal{F}$ as follows (see [23] for a rigorous mathematical setting)

$$[\mathcal{F}(u, \xi)](x, t) := \left[\tilde{\mathcal{F}}(u(x, \cdot), \xi(x))\right](t), \qquad \forall t \in [0, T], \text{ a.e. in } \hat{\Omega}. \tag{2}$$

We notice that operator $\tilde{\mathcal{F}}$ is local in space, i.e., the output $[\mathcal{F}(u, \xi)](x, t)$ depends on $u(x, \cdot)|_{[0,t]}$, but not on $u(y, \cdot)|_{[0,t]}$ for $y \neq x$.

The general setting that we have discussed so far, and the majority of the hysteresis models proposed in the literature, are scalar. Thus, they can be applied only to model unidirectional inputs. However, in many applications, the hysteresis is characterized by a vector input $\boldsymbol{u}(t)$ and vector output $\mathcal{F}(\boldsymbol{u}, t)$; thus, vector hysteresis is encountered. This is the case, for instance, of electric devices such as actuators, transformers, or rotating machines in which the direction of the magnetic field is apriori unknown. The properties of vector hysteresis are often very different from the properties of scalar hysteresis, and the derivation of a general model of vector hysteresis remains an open question. Useful references on vector hysteresis models and their features are given in [12, 26–30].

3. The Preisach hysteresis model

The Preisach model [31] is the most common and probably the most important model to represent magnetic hysteresis phenomenona in the literature. It was originally proposed by the physicist F. Preisach [10] in 1935 in the context of ferromagnetism and later the formalism was more broadly generalized to describe hysteretic behaviors in different fields [6, 8]. Nowadays, it is recognized as a fundamental tool for describing hysteretic systems with complex behaviors.

To describe the hysteresis, the classical Preisach model assumes that each particle of the material has an associated elementary hysteresis operator, called *relay operator*, which characterizes its state. Thus, the system can be modeled as a sum of elementary relays, whose calculation can be performed in parallel, weighted by a distribution function μ, called *Preisach density function*, determining, in a certain sense, the local influence of each relay on the global model. Then, the relationship between the input $u(t)$ and the output $w(t)$ is given by means of the integral:

$$w(t) = \iint_{\rho_2 \geq \rho_1} \gamma_{\rho_1 \rho_2}(u(t), \xi) \mu(\rho_1, \rho_2)\, \mathrm{d}\rho_1\, \mathrm{d}\rho_2, \tag{3}$$

where μ is the weight function, ξ contains the information about the "initial state" of every point of the domain and $\gamma_{\rho_1\rho_2}$ is the relay. The function μ is a non-negative function with compact support in the Preisach half-plane with $\rho_1 \geq \rho_2$ that identifies the system. In practice, it can be analytically approximated by, e.g., Lorentzian or Gaussian distributions (see [9]) or, as it will be explained in Section 3.3, estimated from experimental measurements using the so-called *Everett function* (see [8]). Depending on the properties of these relays and the relationships between them, two types of Preisach models are defined: the *classical (or static) Preisach model* based on the rate-independent relay, and the *dynamic Preisach model* introduced in [14] based on the dynamic relay.

3.1 The static or rate-independent scalar Preisach operator

In this model, the elementary relay operator is represented by a rectangular loop in the input–output plane with "up" and "down" switching values (see **Figure 2** for an example). Each elemental relay, here denoted by h_ρ, is associated to a point $\rho := (\rho_1, \rho_2) \in \mathbb{R}^2$, with $\rho_1 < \rho_2$, and, for any ρ, it has two states: "up" ($h_\rho = 1$) and "down" ($h_\rho = -1$) and two switching thresholds: ρ_2 is the switch-up threshold, and ρ_1 is the switch-down one.

Formally, for any continuous function $u \in C([0, T])$ and initial condition $\xi \in \{1, -1\}$, $h_\rho(u, \xi)$ is a real function defined in the time interval $[0, T]$ such that,

$$h_\rho(u, \xi)(0) := \begin{cases} -1 & \text{if } u(0) \leq \rho_1, \\ \xi & \text{if } \rho_1 < u(0) < \rho_2, \\ 1 & \text{if } u(0) \geq \rho_2. \end{cases} \tag{4}$$

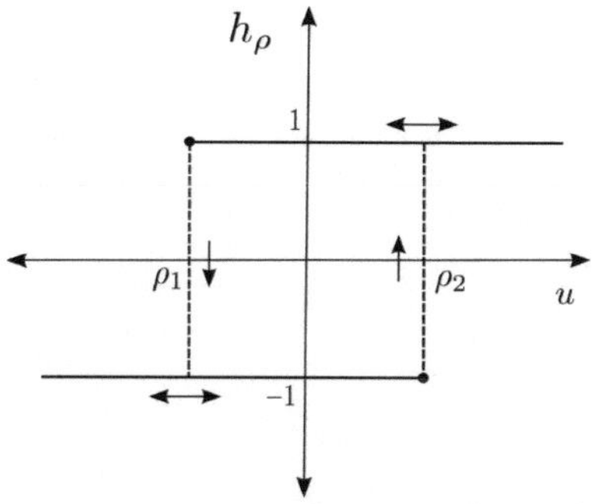

Figure 2.
Scalar relay.

Then, for any $t \in (0, T]$, we consider the set $X_u(t) := \{\tau \in (0, t] : u(\tau) = \rho_1 \; or \; \rho_2\}$. This set keeps account of the previous time instants in which u presents the thresholds ρ_1 or ρ_2. Next, we define

$$h_\rho(u,\xi)(t) := \begin{cases} h_\rho(u,\xi)(0) & \text{if } X_u(t) = \emptyset, \\ -1 & \text{if } X_u(t) \neq \emptyset \quad \text{and} \quad u(\max X_u(t)) = \rho_1, \\ 1 & \text{if } X_u(t) \neq \emptyset \quad \text{and} \quad u(\max X_u(t)) = \rho_2. \end{cases} \tag{5}$$

If $u(0) = 0$, then the following initial condition can be considered

$$h_\rho(u,\xi)(0) := \begin{cases} -1 & \text{if } \rho_1 + \rho_2 > 0, \\ 1 & \text{if } \rho_1 + \rho_2 < 0. \end{cases} \tag{6}$$

When working with ferromagnetic materials, this initial configuration results in zero magnetic induction; thus, the material is often said to be "demagnetized" or in a "virginal" state. We remark that, in this situation, h_ρ can only take values ± 1 and it changes instantaneously from its last value depending on the previous evolution of the system; more precisely, when u reaches the threshold ρ_2 from below, it "switches-up" to value 1, and when it attains ρ_1 from above, it "switches-down" to -1. Therefore, h_ρ is not a local in time mapping: $h_\rho(u, \xi_\rho)(t)$ not only depends on $u(t)$ but on its past history. The *classical Preisach operator* $\mathcal{F}_S$ is then defined as

$$w(t) = [\mathcal{F}_S(u,\xi)](t) = \int_{\mathcal{T}} [h_\rho(u, \xi(\rho))](t) p(\rho) d\rho, \tag{7}$$

where we recall that variable $\xi(\rho)$ contains all the information of the initial state at each point of the system and $p > 0$ denotes the density function.

The integral in (7) is calculated in the so-called *Preisach triangle* (see **Figure 3**, right):

$$\mathcal{T} := \{\rho = (\rho_1, \rho_2) \in \mathbb{R}^2 : -\rho_0 \leq \rho_1 \leq \rho_2 \leq \rho_0\}, \tag{8}$$

being $(-\rho_0, \rho_0)$ the thresholds setting the minimum and maximum values of $u(t)$. By using this triangle, it is possible to provide a geometric interpretation of the Preisach operator. The following is a summary of the most important aspects. For an extended description, the reader can consult Refs. [8, 23]. For a given $u \in C([0, T])$ and any time t, the triangle is split into two sets (one of them possibly empty): $S_u^+(t)$ and $S_u^-(t)$

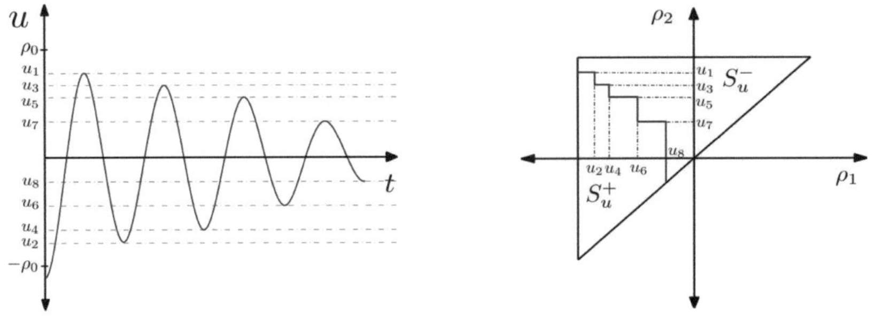

Figure 3.
An arbitrary input $u(t)$ (left) and its corresponding map on the Preisach triangle (right).

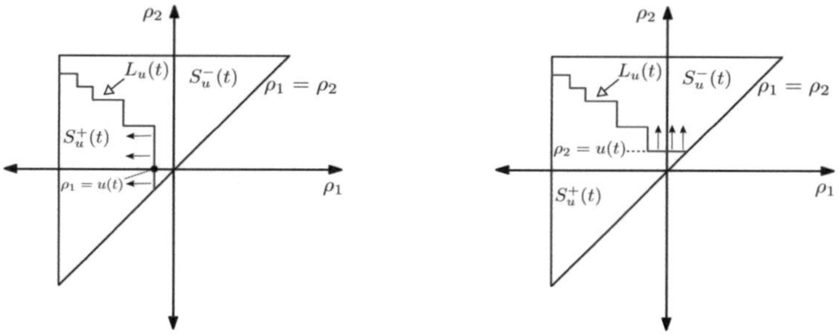

Figure 4.
Staircase line $L_u(t)$ moving right to left (left) and moving up (right).

containing the points (ρ_1, ρ_2) for which the associated relays $h_\rho(u)$ are positive or negative, respectively, i.e.,

$$S_u^-(t) = \{\rho \in \mathcal{T} : [h_\rho(u)](t) = -1\} \quad \text{and} \quad S_u^+(t) = \{\rho \in \mathcal{T} : [h_\rho(u)](t) = 1\} \tag{9}$$

(see **Figure 3**). The interface $L_u(t)$ between the two sets is a staircase line with vertices having coordinates (ρ_1, ρ_2) respectively coinciding with the local minimum and maximum values of u at previous time instants. At time t, $L_u(t)$ intersects the line $\rho_1 = \rho_2$ at $(u(t), u(t))$. $L_u(t)$ moves up as $u(t)$ increases and from right to left as $u(t)$ decreases (see **Figure 4**).

Then, the Preisach operator (7) can be equivalently expressed as

$$w(t) = \int_{S_u^+(t)} p(\rho)d\rho - \int_{S_u^-(t)} p(\rho)d\rho = 2\int_{S_u^+(t)} p(\rho)d\rho - \int_{\mathcal{T}} p(\rho)d\rho, \tag{10}$$

which can be readily calculated once $S_u^+(t)$ is obtained. Since the second integral in (10) is constant in time, the computation of the output $w(t)$ depends on the effective computation of the first integral.

From (10), is clear that $w(t)$ depends on the shape of the interface $L_u(t)$ and the latter is determined by the extremal values of $u(t)$, at previous instants of time. Notice that not all extremal input values are needed. In fact, considering the dependence on $L_u(t)$, the Preisach operator exhibits a *wiping-out property*. This means that every time the input reaches a local maximum $u(t)$, $L_u(t)$ erases ("wipes out") the previous vertices of the staircase having a ρ_2 value lower than the value $u(t)$. Similarly, when an input reaches a local minimum $u(t)$, the memory curve wipes out all previous vertices with a ρ_1 greater than the value of $u(t)$. Therefore, only dominant extreme values contribute to the model, while all the other local extreme of the input are eliminated. **Figure 3** illustrates this in a particular case.

From the description above, we conclude that three basic steps characterize the model application:

- Identification, from experimental data, of the Preisach density p or a suitable auxiliary function.
- At each time step the set $S_u^+(t)$ has to be characterized from the local maxima and minima of the input u. This allows us to update the memory state of the system.

- The value of $w(t)$ is obtained by computing the integral of p in the domain $S_u^+(t)$.

The first and third steps, namely, the identification of the Preisach function and the output calculation are difficult issues. To obtain an efficient procedure for the computation of $w(t)$, we use, as usual, the so-called Everett function identified with the First-Order Reversal Curves (FORC).

3.2 The dynamic or rate-dependent Preisach operator

In [14, 32], Bertotti introduces a rate-dependent generalization of the classical Preisach model, aiming to take into account the rate of change of the input u. This new operator, termed as *dynamic* Preisach operator, overcomes some limitations of the classical model, in particular, the fact that the frequency of the input was not reflected in the shape of the hysteresis diagram.

Contrary to the classical relay operator, where only the two states ± 1 are possible, the dynamic relay can attain all the intermediate values in the interval $[-1, 1]$, switching at a finite velocity which is assumed to be proportional to the difference between $u(t)$ and the threshold values ρ_1 and ρ_2 (see **Figure 5**, right panel). The proportionality factor is a parameter k depending on the material.

In a formal manner, and inspired by [14], for a fixed $\rho = (\rho_1, \rho_2) \in \mathbb{R}^2$, $\rho_1 < \rho_2$, we define the dynamic relay operator η_ρ such that, for any u and $\xi \in [-1, 1]$, $\eta_\rho(u, \xi) : [0, T] \to [-1, 1]$ is the unique function y such that $y(t) \in [-1, 1]$ be the solution of the (nonlinear) Cauchy problem:

$$\frac{dy}{dt}(t) = F(t, y(t)) := \begin{cases} k(u(t) - \rho_2)^+ \quad -k(u(t) - \rho_1)^- & \text{if } -1 < y(t) < 1, \\ 0 & \text{if } y(t) = -1 \text{ and } u(t) \le \rho_2, \\ k(u(t) - \rho_2) & \text{if } y(t) = -1 \text{ and } u(t) \ge \rho_2, \\ k(u(t) - \rho_1) & \text{if } y(t) = 1 \text{ and } u(t) \le \rho_1, \\ 0 & \text{if } y(t) = 1 \text{ and } u(t) \ge \rho_1, \end{cases} \tag{11}$$

$$y(0) = \xi. \tag{12}$$

In the expressions above, the standard notations $x^+ = \max\{x, 0\}$ and $x^- = \max\{-x, 0\}$, have been used; thus $x = x^+ - x^-$. We remark that in this dynamic model the initial state ξ can attain all the values in the interval $[-1, 1]$. Moreover,

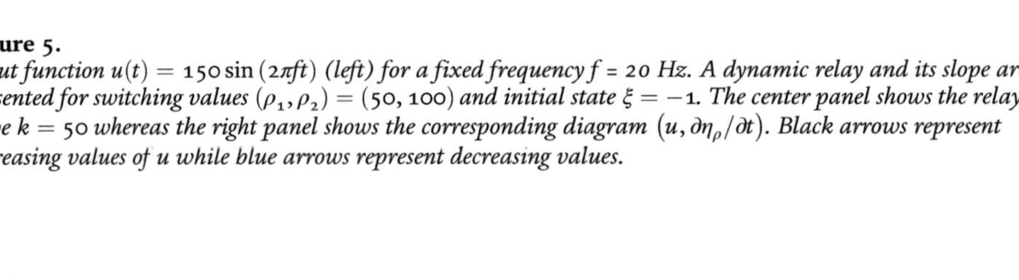

Figure 5.
Input function $u(t) = 150 \sin(2\pi f t)$ (left) for a fixed frequency f = 20 Hz. A dynamic relay and its slope are presented for switching values $(\rho_1, \rho_2) = (50, 100)$ and initial state $\xi = -1$. The center panel shows the relay with slope $k = 50$ whereas the right panel shows the corresponding diagram $(u, \partial\eta_\rho/\partial t)$. Black arrows represent increasing values of u while blue arrows represent decreasing values.

definition (11) is consistent with that given in [2] because cases third and fourth in (11) cannot occur. Nevertheless, if one wants to perform a proper mathematical analysis, all cases must be considered. This mathematical analysis is out of the scope of the chapter; the interested readers are referred to reference [17].

Next, we consider two examples aiming to give an idea of the dynamic curve $(u(t), y(t))$. For the first one, illustrated in **Figure 5**, a sinusoidal input $u(t) = 150\sin(2\pi ft)$ is considered for frequency $f = 20$ Hz and initial condition $\xi = -1$. This figure also shows the curve $(u(t), y(t))$ parameterized with respect to the time variable when the relay η_ρ is characterized by switching values $(\rho_1, \rho_2) = (50, 100)$ and slope $k = 50$. The right panel shows the slope $\partial\eta_\rho/\partial t$, which represents Eq. (11). From the diagram it can be seen that when η_ρ reaches value 1 or -1, the derivative is again equal to zero, as in the interval between switching values (ρ_1, ρ_2). The second example, illustrated in **Figure 6**, shows the dynamic relay when different frequencies of the input and slopes k are considered. From the center panel, it can be seen that the dynamic relay is rate-dependent. On the other and, the right panel shows curve $(u(t), y(t))$ for slopes $k = 10^2$ (dash-dotted line), $k = 10^4$ (dashed line) and $k = 10^8$ (solid line). Notice that the solid line is an approximation to the discontinuous static relay of the classical Preisach model (see **Figure 2**).

From previous considerations, the dynamic Preisach operator $\mathcal{F}_D$ can be defined as

$$w(t) = [\mathcal{F}_D(u, \xi)](t) = \int_T [\eta_\rho(u, \xi(\rho))](t)p(\rho)\,d\rho. \tag{13}$$

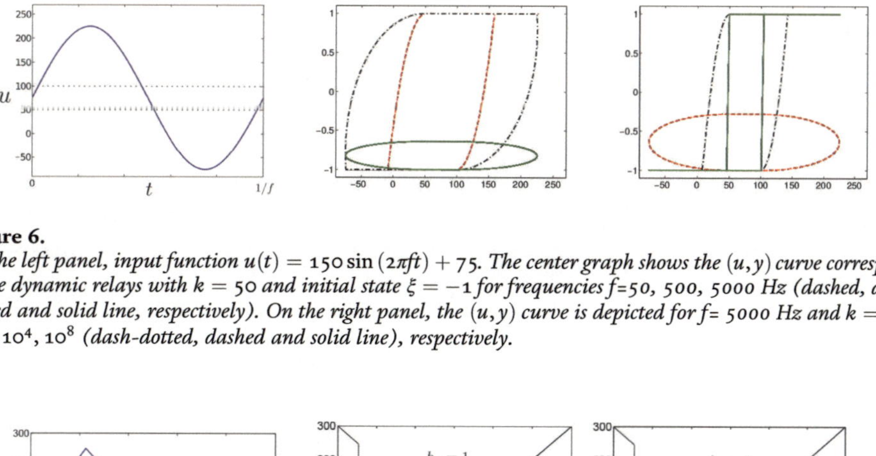

Figure 6.
On the left panel, input function $u(t) = 150\sin(2\pi ft) + 75$. The center graph shows the (u, y) curve corresponding to the dynamic relays with $k = 50$ and initial state $\xi = -1$ for frequencies f=50, 500, 5000 Hz (dashed, dash-dotted and solid line, respectively). On the right panel, the (u, y) curve is depicted for f= 5000 Hz and $k = 10^2, 10^4, 10^8$ (dash-dotted, dashed and solid line), respectively.

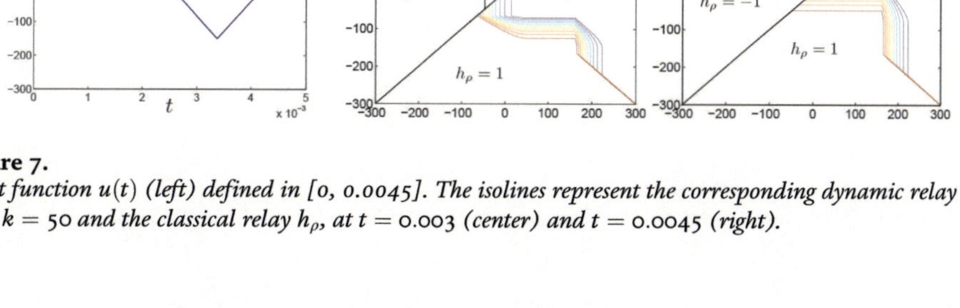

Figure 7.
Input function $u(t)$ (left) defined in [0, 0.0045]. The isolines represent the corresponding dynamic relay values η_ρ with $k = 50$ and the classical relay h_ρ, at t = 0.003 (center) and t = 0.0045 (right).

Notice that, if η_ρ is replaced by h_ρ the classical, rate-independent, Preisach model is obtained. **Figure 7** shows the classical and dynamic relay configurations with respect to the input u. The Preisach triangle is characterized by $\rho_0 = 300$, a constant $k = 50$ and the demagnetized state as initial condition.

3.3 Computation of the Preisach operator based on Everett function

In [8], Mayergoyz developed an approach for the computation of the Preisach model that does not require the Preisach density function p but *the Everett function*, which describes the effect of p on the hysteresis operator. The Everett function is obtained from the First-Order Reversal curves by a procedure described below. A FORC diagram is generated from a class of minor hysteresis loops referred to as first-order reversal curves. A FORC branch $F_{\rho_2'}$ is associated to the threshold ρ_2'. The input u rises up from the "reset" state (every relay is in the "down" state, i.e., $S_u^-(t) = \mathcal{T}$ and $u = -\rho_0$). The output value in the ρ_2' point is called $w_{\rho_2'}$ (inversion point). Then, u is brought back to $-\rho_0$. The branch $F_{\rho_{2'}}$ is drawn by taking the output value $w_{\rho_1',\rho_2'}$ for any value $u = \rho_{1'}$ as is shown in **Figure 8** (left). The branch ends for $u = -\rho_0$, when $S_u^-(t) = \mathcal{T}$. The Everett function is defined in [8] as.

$$E(\rho_1', \rho_2') := \frac{w_{\rho_2'} - w_{\rho_2',\rho_1'}}{2}, \tag{14}$$

which is half of the output variation along the FORC branch starting in ρ_2'. The Everett function and the Preisach function are related by the following integral:

$$E(\rho_1, \rho_2) = \int_{\mathcal{T}(\rho_1,\rho_2)} p(\rho)\, d\rho \qquad \forall(\rho_1, \rho_2) \in \mathcal{T}, \tag{15}$$

where the integration domain $\mathcal{T}(\rho_1, \rho_2)$ is the triangle highlighted in **Figure 8** (right). It should be noted that the integral on the triangle $\mathcal{T}(\rho_1, \rho_2)$ is a function of its upper-left corner and that $E(\rho_1, \rho_2) = 0$ if $\rho_1 = \rho_2$ ($\mathcal{T}(\rho_1, \rho_2)$ degenerates in a single point). The introduction of the Everett integral simplifies the computation of the first integral on the right hand side of (10). First of all, we subdivide $S_u^+(t)$ into k trapezoids $Q_k(t)$ so that $S_u^+ = \cup_k Q_k$. Moreover, each trapezoid can be represented as the set difference of two triangles $\mathcal{T}(m_{k-1}, M_k)$ and $\mathcal{T}(m_k, M_k)$:

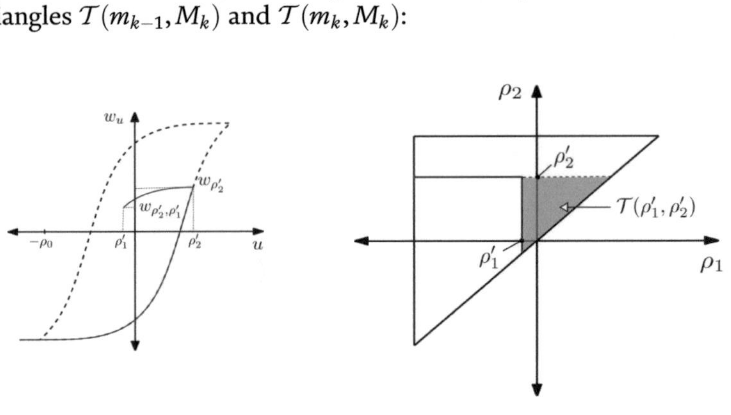

Figure 8.
Left – First order reversal curves (solid line) and major loop (dashed line). Right – Triangle $\mathcal{T}(\rho_1, \rho_2)$.

$$\int_{Q_k(t)} p(\rho)d\rho = \int_{\mathcal{T}(m_{k-1},M_k)} p(\rho)d\rho - \int_{\mathcal{T}(m_k,M_k)} p(\rho)d\rho. \tag{16}$$

Taking into account that the integral on a generic triangle $\mathcal{T}(\rho_1, \rho_2)$ can be expressed by (15), the Preisach integral (10) becomes

$$w(t) = 2\sum_{k=1}(E(m_{k-1}, M_k) - E(m_k, M_k)) - E(-\rho_0, \rho_0), \tag{17}$$

where $E(-\rho_0, \rho_0) = \int_{\mathcal{T}} p(\rho)d\rho$. Form the previous procedure we obtain the following results. First, the output can be computed by using a simple linear combination of the E values in the memory state points represented by the vertices on S_u^+. Second, the value of the Preisach density p is not required, since the identification of E in the domain $\mathcal{T}$ is sufficient to apply the model. The identification of E is simple and it is defined by a repeatable and reliable procedure based on experimental data (FORC branch).

4. Some applications of the hysteresis modeling in electrical engineering

In this section we present some applications of hysteresis modeling in electrical engineering. The first two are motivated by the hysteresis observed when modeling the BH curve of a ferromagnetic material. The third one has to do with the hysteresis present in the State of Charge (SoC)-Open Circuit Voltage (OCV) mapping of Li-ion batteries.

4.1 Hysteresis power losses computation

For the performance analysis of electrical machines, an important factor to be considered is the power losses that occur in the ferromagnetic materials that make up the core of the machine. These losses, usually known as iron losses, can generally be divided into eddy current, hysteresis and excess losses. Eddy current losses are resistive losses due to the currents induced in the magnetic material by the time varying magnetic induction; its magnitude strongly depend on the size of the continuous conductive regions. That is why the core of the machine is usually assembled from thin steel sheets, insulated from each other by a non-conductive coating on the surface.

The hysteresis and excess losses are essentially related to the microscopic properties of ferromagnetic materials. These consist of small magnetic domains, which tend to align along the external (time – varying) applied magnetic field. As a consequence of the molecular friction produced in this continuous movement, heat losses, generally referred to as magnetic hysteresis losses, are produced. Given a material point x, the density of hysteresis losses in the time interval $[t_1, t_2]$ can be computed from the integral:

$$\int_{t_1}^{t_2} \frac{\partial \mathbf{B}}{\partial t}(x,t) \cdot \mathbf{H}(x,t)dt. \tag{18}$$

Nevertheless, these losses are especially difficult to compute, since for ferromagnetic materials, the magnetic induction in each point depends on the intensity of the present magnetic field to which it is exposed, and also on previous exposures to magnetic field intensity of each volume element. This causes differences in the magnetization curve under increasing and decreasing fields; therefore, hysteresis loops arise as the one illustrated in **Figure 9** (left).

In the literature there are several simplified analytical expressions that are used to approximate the different components of the losses, such as those proposed by Bertotti [14], which are among the most widely used. However, the assumptions under which it is possible to apply these formulas are not met in most practical situations. In this context, numerical simulation is a viable option to overcome these limitations.

As an example, we consider an application consisting of the computation of hysteresis and eddy current losses in a laminated medium with toroidal geometry, as sketched in **Figure 9**. It consists of N circular sheets of rectangular section of thickness d surrounded by a coil. We denote by R_1 and R_2 the internal and external radius of the core, respectively. We will assume that the coil is infinitely thin so that it can be modeled as a surface current density (A/m).

As shown in [24], in this situation, it is possible to reduce the computational domain to the meridian section of one single sheet for which it is necessary to derive appropriate boundary conditions. More precisely, the axisymmetric transient eddy current problem to be solved reads:

Find $H(r,z,t)$ and $B(r,z,t)$ such that

$$\frac{\partial B}{\partial t} - \frac{\partial}{\partial r}\left(\frac{1}{\sigma r}\frac{\partial(rH)}{\partial r}\right) - \frac{\partial}{\partial z}\left(\frac{1}{\sigma}\frac{\partial H}{\partial z}\right) = 0 \quad \text{in } \Omega \times (0,T), \tag{19}$$

$$B(r,z,t) = \mu_0(H(r,z,t) + [\mathcal{F}_S(H,\xi)](r,z,t)) \quad \text{in } \Omega \times (0,T), \tag{20}$$

$$H(r,z,t) = h(r,z,t) \quad \text{on } \Gamma \times (0,T), \tag{21}$$

$$B(r,t,0) = B_0(r,z) \quad \text{in } \Omega, \tag{22}$$

where $\Omega = [R_1, R_2] \times [0, d]$, Γ its boundary, and $\sigma(r,z,t), h(r,z,t), \xi(r,z)$ and $B_0(r,z)$ are given functions. We have employed the usual notation: H is the magnetic field, B the magnetic induction and σ the electrical conductivity. Moreover, it can be shown that (see [24] for details):

$$h(r,z,t) := \frac{n_e I(t)}{2\pi r}, \tag{23}$$

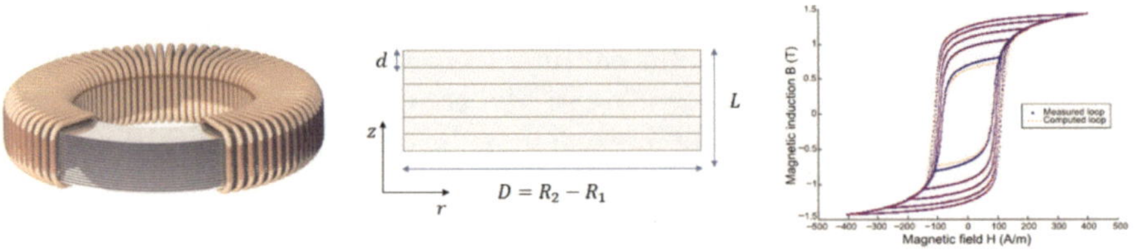

Figure 9.
Toroidal laminated media (left) and its meridian section (center); hysteresis loop measured and approached (right).

where $I(t)$ is the current intensity flowing through the coil at time t and n_e denotes the number of winding turns. In order to model the B-H relation, the classical Preisach hysteresis operator was employed. The Everett function was approximated from experimental data. **Figure 9** (right) shows a comparison between the B-H hysteresis curves measured experimentally and those computed from the Everett function thus assessing the validity of the approximation.

The energy losses can be evaluated by computing the magnetic field and magnetic induction solutions to (19)–(22) with appropriate numerical techniques, such as Finite-Difference Time-Domain (FDTD) [33, 34] or Finite-Element method (FEM) (see, for instance [35–42] where scalar and vector hysteresis models are considered, respectively). In particular, we have applied the latter and a fixed-point iteration (see [24], Section 3) and also [43, 44] for different fixed point iteratios applied to hysteresis models).

In order to compute the losses, we consider the energy balance in the axisymmetric setting, which reads [24]:

$$\underbrace{\int_{\Omega}\left(\int_{t_1}^{t_2}\frac{\partial B}{\partial t}H dt\right) r\, drdz}_{\mathcal{L}_H} + \underbrace{\int_{t_1}^{t_2}\left(\int_{\Omega}\frac{|\mathbf{curl}(H\mathbf{e}_\theta)|^2}{\sigma} r\, drdz\right)dt}_{\mathcal{L}_E} = \underbrace{\int_{t_1}^{t_2}\left((rH)|_{\Gamma}\frac{\partial}{\partial t}\left(\int_{\Omega} B\, drdz\right)\right)dt.}_{\mathcal{L}_T} \quad (24)$$

In this expression, the terms $\mathcal{L}_E$, $\mathcal{L}_H$ and $\mathcal{L}_T$ represent the eddy current, the hysteresis and the total losses, respectively. From the numerical point of view, these losses have been computed (in $(\mathrm{J/m^3})$) as

$$\mathcal{L}_E^h := \frac{1}{\pi d\left(R_2^2 - R_1^2\right)}\left\{\int_0^T\left(2\pi\int_{\Omega}\frac{|\mathrm{curl}(H_h\mathbf{e}_\theta)|^2}{\sigma} r drdz\right)dt\right\}, \quad (25)$$

$$\mathcal{L}_H^h := \frac{1}{\pi d\left(R_2^2 - R_1^2\right)}\left\{2\pi\int_{\Omega}\left(\int_0^T\frac{\partial B_h}{\partial t}H_h dt\right) r drdz\right\}, \quad (26)$$

f(Hz)	B_m(T)	$\mathcal{L}_E^h$	$\mathcal{L}_H^h$	$\mathcal{L}_E^h + \mathcal{L}_H^h$	$\mathcal{L}_T^h$	Total (exp.)[a]	Error 1(%)[b]	Error 2(%)[c]
25	0.5	9.91	126.23	136.14	134.60	121.26	12.27	11.00
	0.9	32.77	269.50	302.27	300.32	300.25	0.67	0.02
	1.4	91.37	585.73	677.10	676.60	638.93	5.97	5.90
150	0.5	48.25	146.10	194.34	186.49	167.55	15.10	11.30
	0.9	178.53	282.28	460.81	4570.42	459.16	0.36	0.46
	1.4	506.40	588.19	1094.58	1095.70	1090.18	0.40	0.51

[a]*Total losses computed experimentally.*
[b]*Relative error between experimental losses and* $\mathcal{L}_E^h + \mathcal{L}_H^h$.
[c]*Relative error between experimental losses and* $\mathcal{L}_T^h$.

Table 1.
Total losses. Numerical versus experimental results (J/m³).

$$\mathcal{L}_T^h := \frac{1}{\pi d \left(R_2^2 - R_1^2\right)} \left\{ 2\pi \int_0^T R_2 H_h \Big|_\Gamma \left(\frac{d}{dt} \int_\Omega B_h dr dz \right) dt \right\}. \tag{27}$$

The measurements reported in **Table 1** were performed on an Epstein frame considering a material sheet of width 30 mm and thickness 0.5 mm. The sheet is subjected to sinusoidal flux excitation with peak induction levels B_m equal to 0.5, 0.9 and 1.4 T and frequencies f equal to 25 and 150 Hz. For each of these peak levels and frequencies, the physical measurements were the total electromagnetic losses per cycle and unit volume and the magnetic field on the boundary of the sheet.
The following geometrical data were considered to simulate the experimental setting with our axisymmetric model: $R_1 = 100$ m, $R_2 = 100.3$ m, $d = 0.0005$ m, $\sigma = 4064777$ $(\text{Ohm/m})^{-1}$. **Table 1** summarizes the results obtained for different frequencies and magnetic induction peak levels. Let us recall that equality $\mathcal{L}_E + \mathcal{L}_H = \mathcal{L}_T$ holds at the continuous level and thus, the difference among the losses $\mathcal{L}_E^h + \mathcal{L}_H^h$ and $\mathcal{L}_T^h$ (columns 6 and 7 in **Table 1**) is due to numerical approximation of the axisymmetric model.

This table shows that the computed losses are close to the losses measured in the experiments. The largest differences are obtained for $B_m = 0.5$ T and are probably due to the fact that our Everett function is unable to generate accurate B-H cycles for "small" values of B (see **Figure 9**, right), making the numerical approximation of the hysteresis losses more inaccurate. These discrepancies can be explained by the *congruency properties* of the classical Preisach model which states that B-H cycles between two fixed extreme magnetic fields H are independent of the induction level B. To avoid these discrepancies, a Moving Preisach hysteresis model can be applied. In this model the congruency property of minor loops is relaxed and it is expected to reproduce lower symmetric hysteresis loops (see [12]).

4.2 Hysteresis modeling in magnetic inspection particle processes

The technique known as Magnetic Particle Inspection (usually MPI for its acronym in English) is a non-destructive method for detecting flaws located on or near the surface of ferromagnetic parts. It exploits the fact that when a ferromagnetic sample is exposed to the influence of a magnetic field, the induced magnetic flux density accumulates inside the material. Then, if there is a crack, the magnetic field will be distorted, causing local magnetic leakage around the defect. Therefore, if fluorescent magnetic particles are sprayed on the magnetized sheet, they will concentrate on the cracks and produce deposits that are easy to identify under ultraviolet light. The purpose of MPI is to find faults in any direction. Because the breakings are easier to detect when orientated perpendicular to the specimen's magnetic field, it is common to apply the

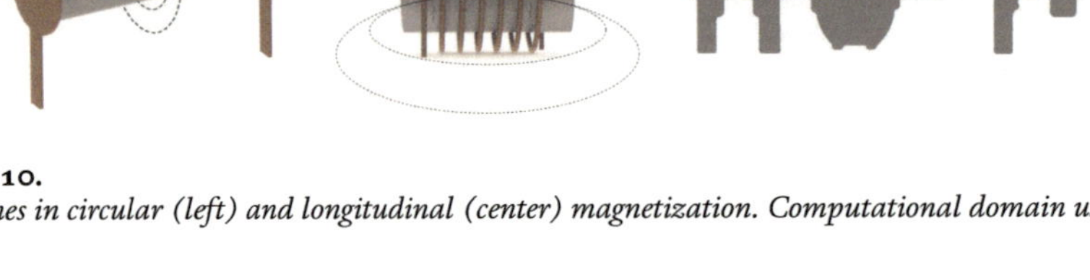

Figure 10.
Field lines in circular (left) and longitudinal (center) magnetization. Computational domain used in tests (right).

magnetization in two orthogonal orientations, such as circular and longitudinal orientation (see **Figure 10**, left). Most times, after inspection, the workpieces must be demagnetized since residual magnetism could interfere with later processing.

In [25], the authors introduced two models for the numerical simulation of the entire magnetization and demagnetization procedures entailed in an MPI test. In particular, the remanent flux density in specimens with cylindrical symmetry is computed. The method is based on scalar models that include magnetic hysteresis and is carried out in three steps: long-term magnetization, circular magnetization and final demagnetization. To achieve circular magnetization (and demagnetization), the workpiece is clamped between two electrical contacts between which an electric current is circulated (see **Figure 10**). For modeling purposes, all fields are assumed to be θ-independent. It is also assumed that the current density has the form $\boldsymbol{J}(\boldsymbol{x},t) \equiv J_z(r,z,t)\boldsymbol{e}_z$, that is, that it traverses the piece along its axial direction. Consequently, the magnetic field stands $\boldsymbol{H}(\boldsymbol{x},t) \equiv H_\theta(r,z,t)\boldsymbol{e}_\theta$. As proven in [25], the problem can be spatially reduced to the meridional section Ω of the workpiece, and reads

$$\frac{\partial \mathcal{F}_S(H_\theta,\xi)}{\partial t}\mathbf{e}_\theta + \mathbf{curl}\left(\frac{1}{\sigma}\mathbf{curl}(H_\theta \mathbf{e}_\theta)\right) = \mathbf{0} \quad \text{in } \Omega \times [t_0, T], \tag{28}$$

$$H_\theta(0,z,t) = 0 \quad \text{on } (0,L) \times [t_0,T], \tag{29}$$

$$H_\theta(R_S(z),z,t) = \frac{I(t)}{2\pi R_S(z)} \quad \text{on } (0,L) \times [t_0,T], \tag{30}$$

$$\frac{\partial H_\theta}{\partial z}(r,z,t) = 0 \quad \text{on } (\Gamma_1 \cup \Gamma_2) \times [t_0,T], \tag{31}$$

where Γ_1, Γ_2 denote the boundaries clamped to the electrical contacts, L is the workpiece length in the axial direction, S_z any cross-section transversal to this direction and $R_S(z)$ its radius. As in previous sections, $\mathcal{F}_S$ is the classical Preisach hysteresis operator and ξ the initial magnetization state.

For the longitudinal magnetization and demagnetization processes, the piece is placed inside a conducting coil carrying an alternating current in the azimuthal direction (see **Figure 10**). To avoid the use of a vector hysteresis law, an infinitely thin conducting surface Γ_S of radius R_S carrying a surface current density $\boldsymbol{J}_S(\boldsymbol{x},t) \equiv J_S(r,t)\boldsymbol{e}_\theta$ is employed to approximate the coil. Let Ω_c be the part to be inspected, R_c representing its radius; moreover, let us denote Ω_0 the air surrounding it which is assumed to be artificially bounded in the r-direction, R_∞ being its outer radius. Then problem reduces to (see [25])

$$\sigma\frac{\partial A_\theta}{\partial t} - \frac{\partial}{\partial r}\left(\mathcal{F}_S^{-1}\left(\frac{1}{r}\frac{\partial}{\partial r}(rA_\theta),\xi\right)\right) = 0 \quad \text{in } (0,R_c) \times [t_0,T], \tag{32}$$

$$\frac{\partial}{\partial r}\left(\frac{1}{\mu_0}\frac{1}{r}\frac{\partial}{\partial r}(rA_\theta)\right) = 0 \quad \text{in } [(R_c,R_S)\cup(R_S,R_\infty)] \times [t_0,T], \tag{33}$$

$$\left[\frac{1}{\mu_0}\frac{1}{r}\frac{\partial}{\partial r}(rA_\theta)\right]_{r=R_S} = J_S \quad \text{on } \{R_S\} \times [t_0,T], \tag{34}$$

$$A_\theta(0,t) = 0 \quad \text{and} \quad \frac{1}{r}\frac{\partial}{\partial r}(rA_\theta)(R_\infty,t) = 0, \forall t \in [t_0,T], \tag{35}$$

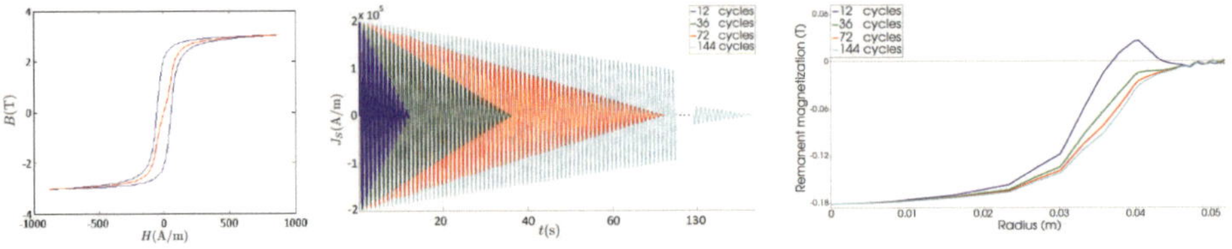

Figure 11.
Left – Major hysteresis loop and (blue) and initial magnetization curve (red). Center – Demagnetizing source. Right – Circular demagnetization. Remanent flux density at $z = 0.2$ m vs. radius. Number of cycles comparison.

where A_θ denotes the azimuthal of component magnetic vector potential. In (34) $[\cdot]_{r=R_S}$ represents the jump across the surface $r = R_S$. Notice that the source is the surface current density $\boldsymbol{J}_S$ which is equal to the jump of $\boldsymbol{H} \times \boldsymbol{n}$ at $r = R_S$.

It is worth noting that when solving (32)–(35), the effective computation of the inverse of the hysteresis operator is avoided by using an iterative algorithm of a fixed point type, based on the properties of the Yosida regularization for maximal monotone operators (see [45]). Moreover, the fields $\boldsymbol{H}$ and $\boldsymbol{B}$ have only one non-null component in the $\mathbf{e}_\theta$ direction in the circular case, and in the $\mathbf{e}_z$ direction in the longitudinal one. This allows using a scalar constitutive law for the nonlinear material, in both the circular and the longitudinal formulations.

For the sake of brevity, only the numerical results corresponding to the circular case are included. The ferromagnetic piece is characterized by the initial magnetization curve and the major hysteresis loop, as shown in **Figure 11**, which was obtained by using an artificial weight function p. The magnetization stage is performed by using a harmonic surface current density on the cylindrical surface Γ_s. When the source is turned off, the workpiece retains a residual field or remanence. In order to remove this remanent field, a surface current density on Γ_s is considered with amplitude equal to the one used for magnetizing the piece and continuously reversing while gradually decreasing to zero (see **Figure 11**). The numerical results give important information about the source needed to successfully demagnetize the piece. In particular, the efficiency of the demagnetization process strongly depends on the source and the number of cycles. In particular, the greater the number of source cycles, the better the demagnetization.

4.3 Hysteresis model applied to battery modeling

It is well-known that hysteresis has a significant impact on the ability to monitor battery performance since it affects the voltage during charging and discharging cycles for different battery chemistries. Therefore, it must be considered for diagnosis algorithms that use the so-called open circuit voltage (OCV) as a measure of the state of charge (SoC) of the battery. SoC indicates the residual charge of the battery with respect to its maximum nominal and it is usually expressed as a percentage of the battery capacity. Since the SoC provides a measure of the available energy of the battery, as well as of its instantaneous power capacity, an accurate estimate is important to monitor battery performance. Unfortunately, it cannot be directly measured but must be estimated from other battery quantities (see, for instance [46–50], to mention some recent works). However, the non-linear electrochemical reactions involved in the

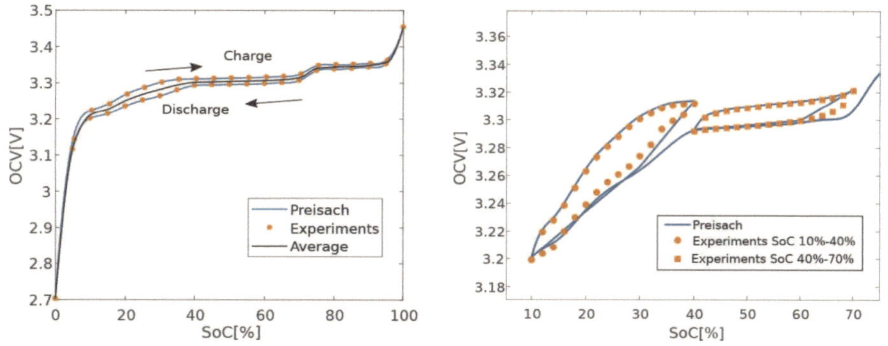

Figure 12.
Measurements along the major loop and the approximation given by the Preisach model (left). Experimental inner loops and approximation of the Preisach model for different SoC values (right).

battery system, the effects of temperature, aging and, specially, hysteresis effects make this task especially difficult.

For a battery cell, hysteresis means that the battery cell reaches different OCV values at the same SoC value and temperature between charge and discharge, depending on the previous charge–discharge history. In particular, from experiments it is shown that, after discharging, the OCV always relaxes below the OCV after charging for the same SoC (see **Figure 12**, left). As a consequence, the SoC-OCV relation is a bundle of curves enclosed in a major loop. The loop consists of two SoC-OCV curves obtained by fully charging and discharging the battery, first to minimum voltage and then to maximum voltage while recording the battery voltage and accumulated ampere hours, for a complete battery cycle. Minor loops lying inside the major SoC-OCV loop can be obtained by conducting similar experiments but with partial cycles (see **Figure 12**, right). A simple approach is to compute the average of the major loop and to relate the battery's OCV with its SoC through this average single-valued curve (see **Figure 12**, left), but usually this approximation suffers from significant errors in the real SoC estimation.

Based on the work [51], in this section we consider a Preisach-type hysteresis model to approximate the SoC-OCV hysteresis loop (see [52–54] for different approaches to handle hysteresis in batteries). The battery dynamics is modeled by using an equivalent circuit model (ECM) as extensively used in the literature [55]. In particular, we consider a general j-th order circuit (see **Figure 13** (left) for the case $J = 2$). It consists of the so-called equivalent series resistance and a number of J resistor-capacitor (R_j, C_j) elements in series. The latter allow us the modeling of the diffusion voltages and are a good approximation of the so-called Warburg impedance. The notations employed in the sequel are the usual ones: Q is the maximum charge (or capacity) of the battery, $i(t)$ is the current intensity, $\eta(t)$ the Coulombic efficiency,

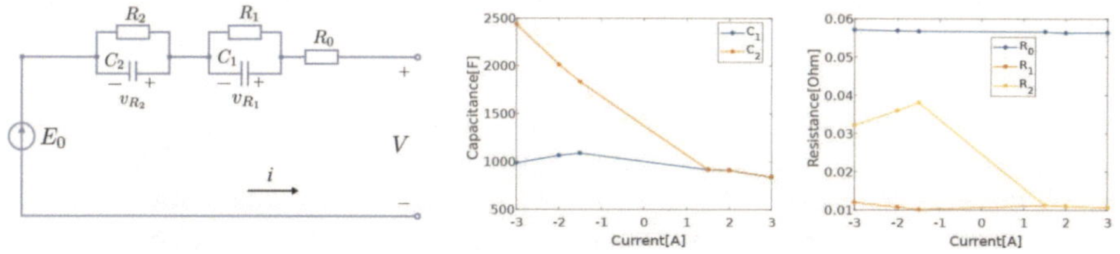

Figure 13.
Sketch of the 2RC model and parameters values.

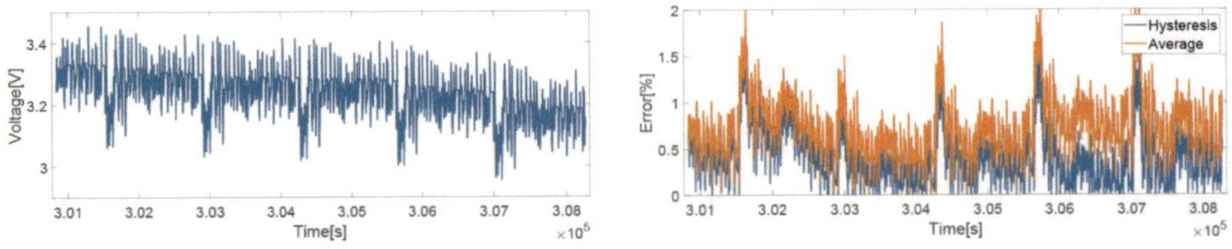

Figure 14.
Percentage error in the voltage approximation (left). Measured and computed voltage with ECM and Preisach model (right).

$V(t)$ the voltage, $z(t)$ the SoC, R_j the $j-$th polarization resistance, C_j the $j-$th polarization capacitance, and $v_{R_j}(t)$ is the voltage between the ends of the the j-th resistor R_j. Thus, the problem to solve reads:

Given functions $\widehat{OCV}$, $i(t)$ and $\eta(t)$, constant parameters Q, R_0, R_j and C_j, and initial conditions z_0 and $v_{R_{j,0}}$, $j = 1, \ldots, J$, find functions $z(t)$, $V(t)$ and $v_{R_j}(t)$, satisfying,

$$\frac{dz}{dt}(t) = -\frac{1}{Q}\eta(t)i(t) \tag{36}$$

$$\frac{dv_{R_j}}{dt}(t) + \frac{1}{R_j C_j}v_{R_j}(t) = \frac{1}{C_j}i(t), \quad j = 1, \ldots, J \tag{37}$$

$$v_{R_j}(0) = v_{R_{j,0}} \quad j = 1, \ldots, J \tag{38}$$

$$z(0) = z_0 \tag{39}$$

$$V(t) = \widehat{OCV}(z(t)) - \sum_{j=1}^{J} v_{R_j}(t) - R_0 i(t) \tag{40}$$

where function $\widehat{OCV}$ gives the OCV from the state of charge z.

From a macroscopic standpoint, the SoC-OCV hysteresis relation can be considered rate-independent; this implies that the OCV depends on the SoC history and not on the velocity (battery current rate) of SoC. Our aim is now to apply a hysteresis model to a lithium-iron-phosphate (LFP) cell. It is known that the classical Preisach operator cannot be directly applied to model the SoC-OCV hysteresis loop. Thus, the hysteretic behavior is modeled with a modified Preisach operator in which the SoC is assumed to be the independent variable. In particular we consider

$$\widehat{OCV}(z(t)) = \int_{S_z^+(t)} p(\rho)d\rho + OCV_{min} \tag{41}$$

where S_z^+ is defined in (9) and OCV_{min} is the minimum OCV value. As usual, the computation of the hysteretic term in (41) is done by means of the Everett function instead of the Preisach hysteresis function p. This function is identified from experimental data considering only the charging SoC-OCV curves of the battery, which is less time consuming than other approaches proposed in the literature. The SoC-OCV curves computed with the Preisach model (41) are compared with experimental data. The approximations are reported in **Figure 12**, where the experimental data and values simulated with the Preisach model are plotted. From this figure, it can be seen that the model approximates the measurements of both the major loop and the inner loops.

Next, we consider a realistic vehicle driving profile to quantify the accuracy of the ECM with hysteresis. We have solved (36)–(39) and compared the computed voltage (40) with the experimental values. The values of OCV are obtained with the average open circuit voltage curve (see **Figure 12**, left) and with the Preisach hysteresis model (cf. (41)). **Figure 14** shows the results obtained when a second-order circuit model (2RC) is considered. More precisely, this figure shows, on the left, the measured voltage and on the right the error between the experimental voltage and the voltage computed with the average curve (in red) and the Preisach model (in blue), respectively. The result has been excellent when the Preisach model is considered, having a relative error of 0.4%.

5. Conclusions

In this chapter we summarize some applications of the hysteresis Preisach model in electrical engineering. First, we have introduced the classical or rate-independent Preisach hysteresis model and the rate-dependent extension of this model, the so-called dynamic Preisach model. Moreover, the identification of this operator based on the Everett function is explained. Finally, we provide some examples of application of the Preisach model in electrical engineering which highlight the importance of having a good hysteresis model in those processes that are characterized by this complex behavior.

Acknowledgements

Work partially founded by FONDECYT-Chile project 1211030, by Xunta de Galicia project GI-1563 ED431C 2021/15 and by FEDER, Ministerio de Economía, Industria y Competitividad-AEI under grant project MTM2017-86459-R.

Author details

Alfredo Bermúdez[1,2], Dolores Gómez[1,2] and Pablo Venegas[3]*

1 Universidade de Santiago de Compostela, Santiago de Compostela, Spain

2 Galician Center for Mathematical Research and Technology (CITMAga), Galicia, Spain

3 GIMNAP, Universidad del Bío-Bío, Concepción, Chile

*Address all correspondence to: pvenegas@ubiobio.cl

References

[1] J. A. Ewing. X. experimental researches in magnetism. Philos. Trans. R. Soc. Lond., 176:523–640, 1885.

[2] M.A. Krasnosel'skiĭ and A.V. Pokrovskiĭ. Systems with Hysteresis. Springer, Berlin, 1989 (Russian edition, Nauka, Moscow (1983)).

[3] J. W. Macki, P. Nistri, and P. Zecca. Mathematical Models for Hysteresis. SIAM Review, 35(1):94–123, 1993.

[4] V. Hassani, T. Tjahjowidodo, and N. D. Thanh. A survey on hysteresis modeling, identification and control. Mech. Syst. Signal Proc., 49(1-2, SI): 209–233, 2014.

[5] A. Visintin. Chapter 1 – Mathematical Models of Hysteresis. In Giorgio Bertotti and I. D. Mayergoyz, editors, The Science of Hysteresis, pages 1–123. Academic Press, Oxford, 2006.

[6] A. Visintin. Differential models of hysteresis, volume 111 of Applied Mathematical Sciences. Springer-Verlag, Berlin, 1994.

[7] M. Brokate and J. Sprekels. Hysteresis and Phase Transitions. Springer, Berlin, 1996.

[8] I. D. Mayergoyz. Mathematical models of hysteresis. Springer-Verlag, New York, 1991.

[9] G. Bertotti. Hysteresis in Magnetism. Academic Press, 1998.

[10] F. Preisach. Über die magnetische nachwirkung. Zeitschrift für Physik, 94: 277–302, 1935.

[11] D. A. Philips, L. Dupré, J. Cnops, and J. A. A. Melkebeek. The application of the Preisach model in magnetodynamics: theoretical and practical aspects. J. Magn. Magn. Mater., 133(1):540–543, 1994.

[12] E. Della Torre. Magnetic Hysteresis. IEEE Press, New York, 1999.

[13] I. D. Mayergoyz. Dynamic Preisach Model of Hysteresis. IEEE Trans. Magn., 24(6):2925–2927, 1988.

[14] G. Bertotti. Dynamic generalization of the scalar Preisach model of hysteresis. IEEE Trans. Magn., 28(5): 2599–2601, 1992.

[15] Y. Yu, Z. Xiao, N. G. Naganathan, and R. V. Dukkipati. Dynamic Preisach modelling of hysteresis for the piezoceramic actuator system. Mech. Mach. Theory, 37(1):75–89, 2002.

[16] L. Dupré, G. Bertotti, V. Basso, F. Fiorillo, and J.A.A Melkebeek. Generalisation of the dynamic Preisach model toward grain oriented FeSi alloys. Physica B, 275(1):202–206, 2000.

[17] A. Bermúdez, D. Gómez, and P. Venegas. Mathematical analysis and numerical solution of models with dynamic Preisach hysteresis. J. Comput. Appl. Math., 367:112452, 2020.

[18] O. Bottauscio, M. Chiampi, D. Chiarabaglio, and M. Repetto. Preisach-type hysteresis models in magnetic field computation. Physica B, 275(1-3):34–39, 2000.

[19] B. Schweizer. Hysteresis in porous media: Modelling and analysis. Interface Free Bound., 19(3):417–447, 2017.

[20] V. A. Lubarda, D. Šumarac, and D. Krajcinovic. Preisach model and

hysteretic behaviour of ductile materials. European J. Mech. A Solids, 12(4):445–470, 1993.

[21] A. Rao and A.R. Srinivasa. A two species thermodynamic Preisach model for the torsional response of shape memory alloy wires and sprin under superelastic conditions. Int. J. Solids Struct., 50(6):887–898, 2013.

[22] P. Krejčí, J. P. O'Kane, A. Pokrovskii, and D. Rachinskii. Properties of solutions to a class of differential models incorporating Preisach hysteresis operator. Phys. D, 241(22):2010–2028, 2012.

[23] A. Bermúdez, D. Gómez, R. Rodríguez, and P. Venegas. Mathematical analysis and numerical solution of axisymmetric eddy-current problems with Preisach hysteresis model. Rend. Sem. Mat. Universitá e Politecnico Torino, special volume, 72 (1-2):73–117, 2014.

[24] A. Bermúdez, L. Dupré, D. Gómez, and P. Venegas. Electromagnetic computations with Preisach hysteresis model. Finite Elem. Anal. Des., 126:65–74, 2017.

[25] A. Bermúdez, D. Gómez, M. Piñeiro, P. Salgado, and P. Venegas. Numerical simulation of magnetization and demagnetization processes. IEEE Trans. Magn., 53(12):1–6, 2017.

[26] M. Kuczmann. Vector Preisach hysteresis modeling: Measurement, identification and application. Physica B, 406(8):1403–1409, 2011.

[27] M. Kuczmann. Dynamic extension of vector Preisach model. Physica B, 549: 47–52, 2018. 11th International Symposium on Hysteresis Modeling and Micromagnetics (HMM 2017).

[28] I. D. Mayergoyz. Mathematical Models of Hysteresis and Their Applications. Elsevier, 2003.

[29] E. Cardelli and A. Faba. Modelling of vector hysteresis at macromagnetic scale: Open questions and challenges. Physica B, 486:130–137, 2016.

[30] E. Cardelli, E. Della Torre, and A. Faba. A General Vector Hysteresis Operator: Extension to the 3-D Case. IEEE Trans. Magn., 46(12):3990–4000, 2010.

[31] K. Preis. A contribution to eddy current calculations in plane and axisymmetric multiconductor systems. IEEE Trans. Magn., 19:2397–2400, 1983.

[32] G. Bertotti and M. Pasquale. Physical interpretation of induction and frequency dependence of power losses in soft magnetic materials. IEEE Trans. Magn., 28(5):2787–2789, 1992.

[33] F. Bertoncini, F. Beux, E. Cardelli, S. Di Fraia, and B. Tellini. Stable FDITD formulation for electromagnetic field diffusion in soft magnetic materials. IEEE Trans. Magn., 39(3):1681–1684, 2003.

[34] S. Quondam Antonio, A. Faba, H. P. Rimal, and E. Cardelli. On the analysis of the dynamic energy losses in NGO electrical steels under non-sinusoidal polarization waveforms. IEEE Trans. Magn., 56(4):1–15, 2020.

[35] R. Van Keer, L.R. Dupré, and J.A.A. Melkebeek. On a numerical method for 2D magnetic field computations in a lamination with enforced total flux. J Comput. Appl. Math., 72(1):179–191, 1996.

[36] N. Takahashi, S.-H. Miyabara, and K. Fujiwara. Problems in practical finite

element analysis using Preisach hysteresis model. IEEE Trans. Magn., 35 (3):1243–1246, 1999.

[37] E. Fallah and V. Badeli. A new approach for modeling of hysteresis in 2-D time-transient analysis of eddy current using FEM. IEEE Trans. Magn., 53(7):1–14, 2017.

[38] Y. Li, L. Zhu, and J. Zhu. Core loss calculation based on finite-element method with Jiles–Atherton dynamic hysteresis model. IEEE Trans. Magn., 54(3):1–5, 2018.

[39] L. R. Dupré, R. Van Keer, and J. A. A. Melkebeek. Complementary 2-D finite element procedures for the magnetic field analysis using a vector hysteresis model. Int. J. Numer. Meth. Eng., 42(6): 1005–1023, 1998.

[40] J. B. Padilha, P. Kuo-Peng, N. Sadowski, and N. J. Batistela. Vector Hysteresis Model Associated to FEM in a Hysteresis Motor Modeling. IEEE Trans. Magn., 53(6), 2017.

[41] M. Tousignant, F. Sirois, G. Meunier, and C. Guérin. Incorporation of a vector Preisach–Mayergoyz hysteresis model in 3-D finite element analysis. IEEE Trans. Magn., 55(6):1–4, 2019.

[42] C. Guérin, K. Jacques, R. V. Sabariego, P. Dular, C. Geuzaine, and J. Gyselinck. Using a Jiles-Atherton vector hysteresis model for isotropic magnetic materials with the finite element method, Newton-Raphson method, and relaxation procedure. Int. J. Numer. Model El., 30(5):e2189, 2017.

[43] E. Dlala and A. Arkkio. Analysis of the convergence of the fixed-point method used for solving nonlinear rotational magnetic field problems. IEEE Trans. Magn., 44(4):473–478, 2008.

[44] P. Zhou, D. Lin, *C. Lu*, M. Rosu, and D. M. Ionel. An adaptive fixed-point iteration algorithm for finite-element analysis with magnetic hysteresis materials. IEEE Trans. Magn., 53(10):1–5, 2017.

[45] A. Bermúdez and C. Moreno. Duality methods for solving variational inequalities. Comput. Math. Appl., 7:43–58, 1981.

[46] M. A. Hannan, M. S.H. Lipu, A. Hussain, and A. Mohamed. A review of lithium-ion battery state of charge estimation and management system in electric vehicle applications: Challenges and recommendations. Renew. Sust. Energ. Rev., 78:834–854, 2017.

[47] J. P. Rivera-Barrera, N. Muñoz Galeano, and H. O. Sarmiento-Maldonado. SOC estimation for lithium-ion batteries: Review and future challenges. Electronics, 6(4), 2017.

[48] R. Zhang, B. Xia, B. Li, L. Cao, Y. Lai, W. Zheng, H. Wang, and W. Wang. State of the art of lithium-ion battery SOC estimation for electrical vehicles. Energies, 11(7), 2018.

[49] M. U. Ali, A. Zafar, S. H. Nengroo, S. Hussain, M. Junaid Alvi, and H.J. Kim. Towards a smarter battery management system for electric vehicle applications: A critical review of lithium-ion battery state of charge estimation. Energies, 12 (3), 2019.

[50] Y. Wang, J. Tian, Z. Sun, L. Wang, R. Xu, M. Li, and Z. Chen. A comprehensive review of battery modeling and state estimation approaches for advanced battery management systems. Renew. Sust. Energ. Rev., 131:110015, 2020.

[51] P. Venegas, D. Gómez, M. Arrinda, M. Oyarbide, H. Macicior, and A.

Bermúdez. Kalman filter and classical Preisach hysteresis model applied to the state of charge battery estimation. Submitted.

[52] V. Franzitta, A. Viola, and M. Trapanese. Description of Hysteresis in Lithium Battery by Classical Preisach Model. Adv. Mat. Res., 622-623:1099–1103, 2013.

[53] F. Baronti, N. Femia, R. Saletti, C. Visone, and W. Zamboni. Hysteresis modeling in Li-ion batteries. IEEE Trans. Magn., 50(11):1–4, 2014.

[54] L. Zhu, Z. Sun, H. Dai, and X. Wei. A novel modeling methodology of open circuit voltage hysteresis for LiFePO4 batteries based on an adaptive discrete Preisach model. Appl. Energ., 155:91–109, 2015.

[55] G. L. Plett. Battery Management Systems: Volume 1, Battery Modeling. Artech House, Boston, 2015.

Chapter 5

Hydromagnetic Steady Magnetized Plasma Encountered by Voyager in the Interstellar Space

Daniel Benjamín Berdichevsky

Abstract

With the spacecraft Voyager 1 and 2, we observe from the inside the near interstellar medium that is a very strongly magnetized (|**B**| equal or larger than 0.4 nT) medium, which is very dilute. We deal with a medium about 8-orders of magnitude less dense than the best vacuum possible to achieve in the laboratory. For the considered matter densities the plasma energy is quite small at a random equivalent thermal value of no more than 30,000°K, so that a **B**-field of 1/2 nT constitutes a very strong magnetic field. Consequently, the medium's pressure is dominated by the **B**-field. Based on these everyday measurements, we proceed to interpret the nature of the medium, assuming the reasonable, consistent result, we check with observation that its overall property satisfies hydromagnetism (MHD). The intensity of the **B**-field is consistent with remote sensing of the 21 cm split of the e-line of the atomic Hydrogen, and permits to understand the ancient nature of the medium, for example, constituting an example of permanent magnetism, and in this way we infer a few possibilities on the nature of the whole structure of the surrounding local interstellar magnetic cloud and the evolution of the home galaxy.

Keywords: astrophysics, LISM, heliopause, SW sheath, low beta plasmas, diamagnetism, thermodynamic properties, 3-D amorphous crystalline MHD matter of the Langmuir kind

1. Introduction

The medium we refer is evaluated to have a mixture of neutral 90% and plasma 10% with a particle density of about one particle (atoms + ions) per cm^3. This value indicates a dilute electron density of 0.1 per cm^3. Also, the dominant feature Is the strong presence of a steady magnetic field (**B**-field) with a value that varies from about 0.4 to 0.7 nT. This medium has been in astrophysics already monitored remotely well before the arrival of the spacecraft (SC) Voyager 1 and 2 by the observation of 21 cm frequency signal provided by the e-flip in the orientation of its spin. These remote measurements have a localization uncertainty of ±200 astronomic units (AUs), to which the Voyager SC brings a far better knowledge on the strong **B**-field localization through its in situ measurement. (Stone [1], announces that

Voyager 1 entered local interstellar medium (LISM)). The AU is defined by the mean distance of the Earth to the Sun which is equivalent to 150 solar radii (1 solar radius is the distance from the star center to its photosphere and is well-known to be 700,000 km with an uncertainty of possibly 0.1%). The in situ observations are achieved using in SC located magnetic field, radio, energetic particles, plasma instruments, respectively described respectively in [2–6].

In this work, we use both in situ observation by the SC (Sections 2, 3.1, 3.2, and by its combination with studies also performed by the author in similar strong magnetized MHD matter in Sections 3.3 and 3.4). Further, in Section 4, the author extrapolates the findings to astrophysical conditions from the realm solely of remote sensing (e.g. the Hubble telescope, and other more recent similar missions). Conclusions are drawn in Section 5.

2. The local interstellar medium (LISM)

The LISM encloses the plasma region generated by the Sun in what is known as solar wind (SW), plus its shocked plasma region known as the heliosphere sheath. The interstellar medium, which its in-situ known region we call the very LISM is of uncertain origin. Some workers attribute its formation to be the wake of a supernova that occurred millions of years ago. This region is known as the local interstellar molecular cloud, of which **Figure 1** shows a sketch, and an artistic representation is shown in **Figure 2**.

This interstellar molecular cloud is in the path of the solar system, which by now and from hundreds of thousands of years has been immersed in it. The very LISM is then the region of the molecular cloud modified/surrounding the plasma pushing out of the Sun and the LISM region we explore with SC Voyager since the year 2012 with Voyager 1 (V1), and 2018 with Voyager 2 (V2), see, e.g. Berdichevsky [7] on the sheath region before the arrival of the SC to the heliopause.

Figures 1 and **2** identifiers (e.g. representation of the star Sun) are not at scale, but the distances are, and in the bottom right, it is indicated a 'parsec as the reference unit giving an idea of the dimensions involved in the sketch. The Voyager 1, currently at $\sim 7 \cdot 10^{-4}$ parsec, in several years will arrive at 0.001 parsecs from the Sun at their velocity as it moves, escaping the gravitational field attraction of the solar system.

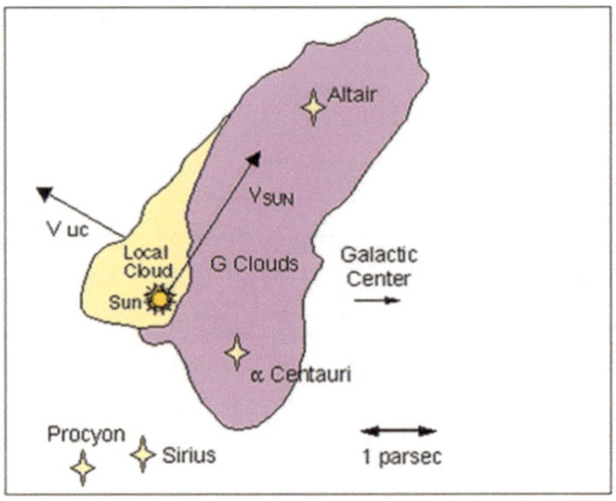

Figure 1.
A sketch, and plane of sky projection of the local molecular cloud including location of the Sun and other stars.

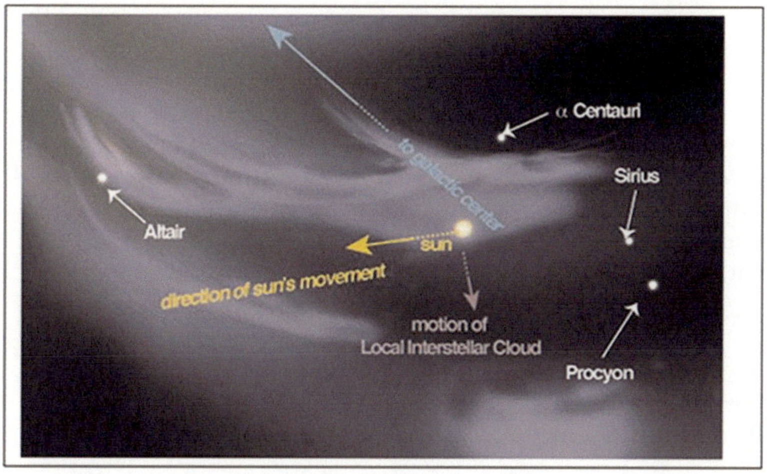

Figure 2.
An artistic representation by Linda huff and Priscilla Frisch.

Observed in the interstellar medium, since its entrance in 2012 by V1, see the announcement by NASA [1], see **Figure 3**, are the large intensity of the **B**-field, and the four times occurrence of a compression process in which changes in **B**-field magnitude coincide with the change of density observed, [8]. Good measurements of the plasma in the LISM have been achieved using the radio instrument in both V1 and V2 [5]. On the other hand, the plasma instrument design to measure properties of the

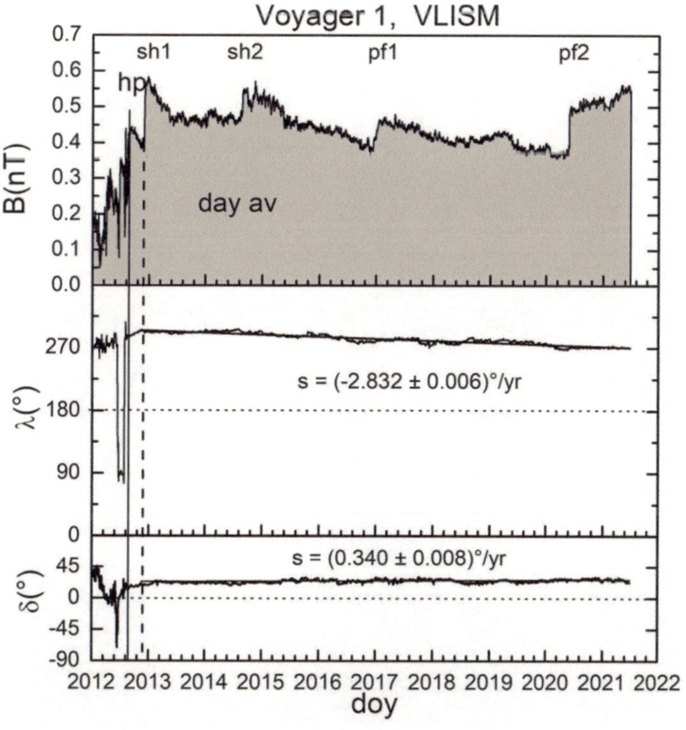

Figure 3.
*Approximately ten years of in situ record by V1 of the **B**-field in the very LISM.*

SW, [3], is not optimal for identifying the LISM plasma properties, but they achieve reasonable evaluations with large errors in V2 (in V1, that instrument failed a long time ago).

The solar system displaces relative to the LISM at solely a few tens of kms^{-1}. At such speed, no shock at the heliopause interface is expected nor it has been encountered. Here, we assume a shock in magnetic field frozen matter (see, e.g. [9]) that had been identified to be consistent with a process called collision-less [10]. However, there are some workers in the field who brought the idea of shocks in the LISM medium that, in their view, are collisional, and that is a current subject of debate (see,

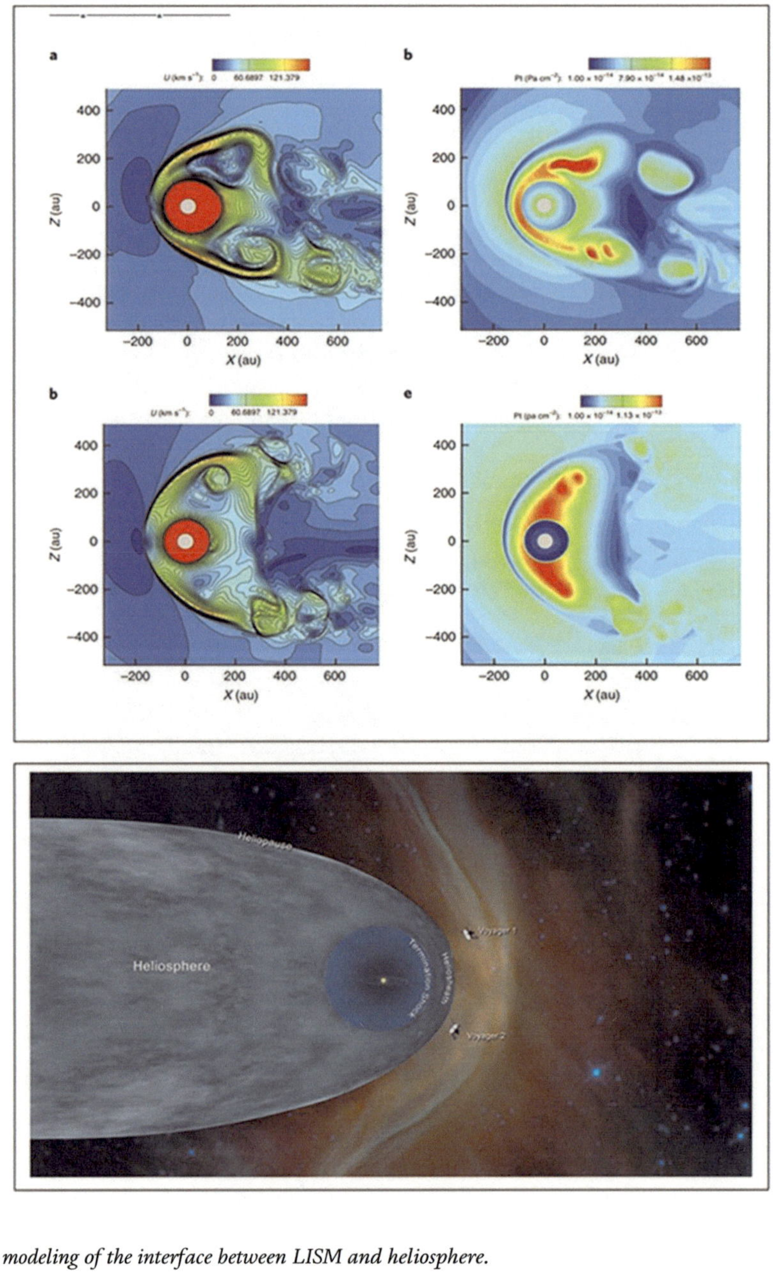

Figure 4.
Two MHD modeling of the interface between LISM and heliosphere.

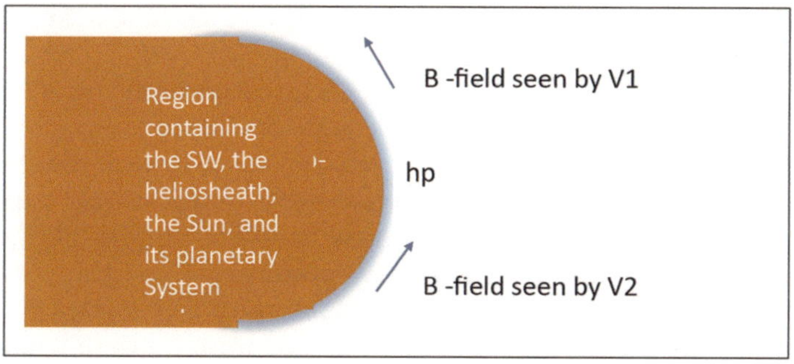

Figure 5.
The orientation of ***B****-field observed both by V1 and V2 outside heliopause (hp).*

e.g. [11]). Then at their suggestions in **Figure 3**, there are compressions labeled as shocks and others just compressions. We will come to the subject of these types of compressions later.

The global representations of the interface between heliosphere and very LISM are shown in **Figure 4**, top panel [12] and bottom panel [13, 14] are competing interpretations which currently dominate the debate on the subject. In the bottom panel, the meridional cut of a 3-D representation is indicated, different main regions of the magnetosphere as well as the location of the V1 and V2 SC relative to a reference astrophysical coordinate system where the top is pointed in a direction which approximately is defined by our heliosphere standard North definition.

Figure 3, in [8], shows that V1 encounters four compression jumps in which density and magnetic field data indicate an equal value for the change, both in the magnitude of the **B**-field as well as in the plasma density content. Based on that fact and also the orientation of the LISM **B**-field wrapped around the heliosphere plasma with its expansion stopped (**Figure 5**) at the heliopause (hp), and which are well identified, we make the educated assumption that the medium possesses the properties of a simple ideal MHD state, see, e.g. [15]. This subject introduces us to the next section.

3. On the MHD nature of the LISM medium

3.1 On the nature of the medium

The supportive arguments for a medium consistent with ideal MHD are the followings:

First, we point out the behavior of magnetic field and plasma density exquisitely agreeing within error uncertainty as observed and listed in **Table 1**, at compressional discontinuities in **Figure 3.**

Event	Date (approx.)	B2/B1	**N2/N1**	N2/N1
sh 2	2014-236	1.13	1.11	1.11
pf1	2017-001	1.19	1.17	1.12
pf2	2020-146	1.35	1.36	1.36

Table 1.
Three examples of compression properties observed in the interstellar medium.

Second, we identify both at V1 and at V2 a **B**-field consistent with a draping of the **B**-field around the magnetopause (**Figure 5**).

Third, the **B**-field shows along the path of V1 (see **Figure 3**) and V2 a steadiness, which is quite characteristic of low plasma pressure and ideally negligible changes. But there are interesting oscillatory changes, which transitioned since the start of V1 immersion in the LISM from mainly compressional to dominatingly transversal self-vibrational modes (see, e.g. [15]) present in low temperature (low plasma pressure) plasmas, plainly known in the field as low beta plasmas.

Fourth, these ripple variations in the **B**-field intensity (see **Figure 6**), occur at sharp compression transition and illustrate there the presence of self-modes (frequency) of the magnetized media up to the resolution of our measurements of the magnetic field (ideal MHD self-mode oscillations).

The conditions above illustrated, [8], indicate a consistency of the medium to be quite close to the ideal MHD. Further, we have a likely existence of these long-lasting magnetized plasma, which indicates that diffusion, loss of the magnetic field has to be negligible and further constitute an argument for a LISM of magnetized matter that satisfies being ideal MHD.

In **Figure 6** (top) are presented observations of distributions of

$$dB_{N1} = B_N(n) – B_N(n-1) \tag{1}$$

and

$$dB_{T1} = B_T(n) – B_T(n-1) \;\text{(oscillations are in the bottom panels in \textbf{Figure 6})} \tag{2}$$

and time series of the 1-hr increments of '*dB_{N1} and dB_{T1}*' from the beginning of the magnetic hump on day 147, 2020 to the end of the magnetic hump. The matter we observe from within a very LISM is extremely dilute see Abstract. For the matter

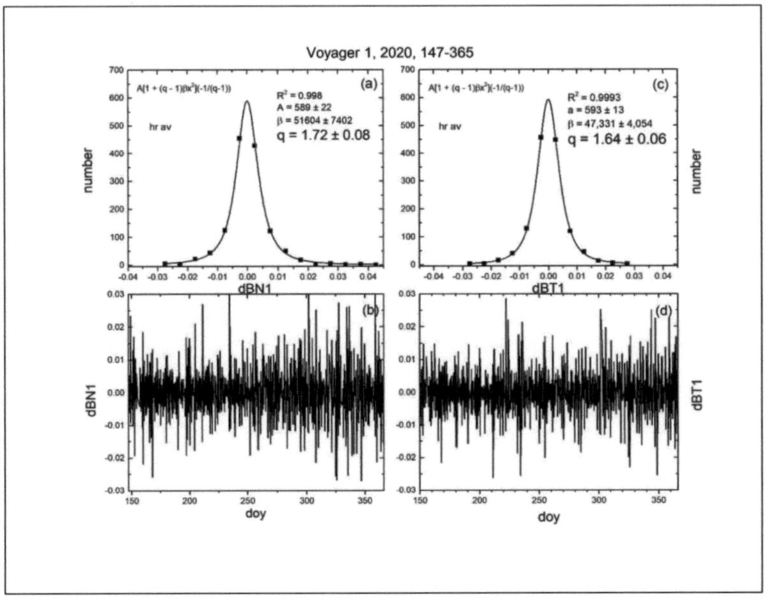

Figure 6.
*Presence of self-oscillation modes, in the **B**-field for DOY 147, the year 2020.*

densities considered, the observation of an average **B**-field close to ½ nT results to be a very strong magnetic field. This quite strong **B**-field can be understood this way when comparing its energy with the kinetic random energy of the matter there present. Other relevant aspects to be considered are the forces acting on this LISM:

- (a) Gravitational self, (b) gravitational with compact objects in its neighborhood, e.g. stars, (c) and gravitational with the galaxy as a whole.

- Magnetic stresses

- Interface stresses with the Sun changing pressures reaching the hp, with other stellar 'winds' possibly enclosed/in-the-local-range of the LISM

The apparent long life of the LISM suggests it is stable, and we assume: ~thermal equilibrium.

Next, we proceed to make some conservative assumptions on size and mass of the local, interstellar molecular cloud (the magnetized plasma structure in which we observe the measurements of the two Voyager SC). From **Figure 1**, let us assume we got a good representation of its base projected on a plane. Hence, we can estimate its size to be:

$$\sim 3 \times 8 \times 8 \text{ parsec}^3 \tag{3}$$

being conservatives and assuming homogeneous mass distribution and rounding the mass estimated to be observed at the Voyager locations, and here the presence of heavier element contributions are included.

$$\sim 2 \text{ nucleons/cm}^3 \tag{4}$$

this gives a mass $\sim 4/100 \cdot 10^{30}$ kg, about 1/50th the mass of the Sun ($- 2 \cdot 10^{30}$ kg).

Next, we ask ourselves the fundamental question: 'What could hold the cloud undisturbed?' and we know that we observe in situ the magnetic field, *actually observed values are consistent with the remote observation from Earth of the splitting of the 21 cm electron-line in atomic H.*

- We also observe plasma, and remember the region of plasma waves (= Plasma-Noise).

- So we think the system to be, by nature magnetohydrodynamic (MHD).
  i.e. a structure that is *superconductor, diamagnetic, super-elastic with unusual properties.*

- Henceforth: the idea of thermodynamics *MHD equilibrium* is strengthened.

Then we proceed to think of the presence of the V1 and V2 SC, in situ observers, in a magnetohydrodynamic steady state. And we further add to the inferences we make that we are in the presence of a magnetized matter with the properties:

- Of a matter medium which possesses a coalescence state, following **Hannes Olof Gösta Alfvén** ideas, e.g. [16]. See, e.g. [17].

- Of a matter medium that moves (90% neutrals and 10% plasma, i.e. neutrals/ions) attached—frozen to the magnetic field.

- We speculate further that such material would have those properties of the one we learn to be familiar with, which are those of a 'solid,' a very porous amorphous crystalline solid with homogeneous domains cross sections of 10^{10} km^2 ... or more ...

Figure 7 represents the outlined view of a charge neutral (i.e. q-neutral) structure. E.g. [18] present conjecture on this amorphous in nature, matter 'atoms/ions' anchored to **B**-field (as a 3-D Langmuir lattice). **B** intensity low to high is represented in the **light** to the **dark blue gradation** of **Figure** 7. The privilege of working with an impressive suite of high-sensitivity instruments in Ulysses and Wind SC provided the opportunity of advancing on the understanding of the constitutive properties of the MHD medium, in its own right, a very stable long-live magnet (permanent magnet) as it is possible to learn about from the WWW archive pages for the Ulysses mission, e.g. https://www.esa.int/Science_Exploration/Space_Science/Sun_set_Ulysses_soarl_mission_on_1_July2, and WWW pages for the currently active Wind SC mission, e.g. h ttps://wind.nasa.gov.

When in a breathing mode process, we have that in small oscillations, the *e-gas* compresses, i.e. pressure increases, a rise that decreases magnetization and the internal energy of the *e-gas* by decreasing '*T*' in a '*diamagnetic*' process (i.e., **B**-field magnitude also decreases). This is a quite subtle effect that the in situ resources available in the Voyager instruments are not capable of observing. In following Section 3.4, we re-visit these understandings and observations.

Nevertheless, features observed in V1 and V2 appear to give the opportunity to learn about the constitutive nature of the diamagnetic permeability of the medium when the observation of the equilibration location at heliopause between heliosphere

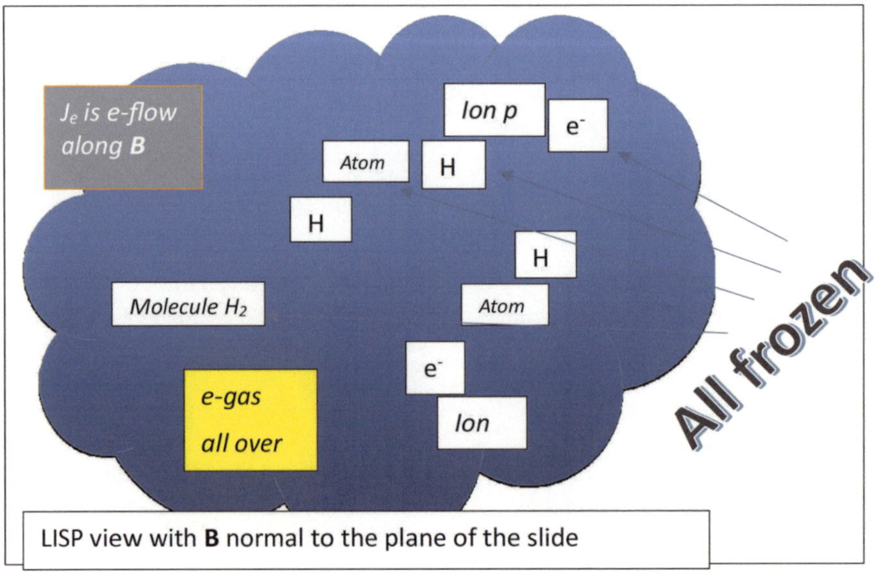

Figure 7.
A representation of the outlined constitutive nature of the LISM environment.

pressure and pressure of the medium surrounding it suggests a pressure larger than the one of the very LISM as it would be essentially provided by a *vacuum* μ_0 of the strong **B**-field observed intensity value of about 0.4 nT at V1.

This value of the **B**-field is too low, assuming the permeability μ_0 of vacuum, and fails to be identified as an equilibration value as it is pointedly indicated in [19]. The understanding that the LIMS medium would have a $\mu \neq \mu_0$-vacuum permeability is made plausible in studies for similar strong **B**-field conditions, which are described in the following Sections 3.3 and 3.4.

3.2 About the conditions on the mass supported by the MHD structure

We may, at this point, idealize the local molecular cloud region as composed of a closely connected set of curved MHD tubes as suggested in the right panel of **Figure 8**, see also the following **Figure 9**. These are expected to be MHD ideal structures (negligible diffusion loss through the millions of years of their existence). Then the stability would be the result between the self-gravitating force of the matter and the magnetic stresses of the magneto-matter state.

Figure 9 shows the simplified MHD structure, tube-shaped with its source currents of its **B**-field which equilibrium conditions are assumed for the interstellar molecular cloud that interacts in its very LISM with the heliosphere region we study in situ with the Voyager SC.

When considering the self-gravitational force of the matter, we obtain along the symmetry axis: $F_x = \nabla - \nabla\phi)_x =$

$$-2\pi\, G N \left\{2x + \left(R^2 + [L/2{-}x]^2\right)^{1/2} - \left(R^2 + [L/2 + x]^2\right)^{1/2}\right\} for{-}L/2 \leq x \leq L/2 \quad (5)$$

$$-2\pi\, G N \left\{L + \left(R^2 + [L/2{-}x]^2\right)^{1/2} - \left(R^2 + [L/2 + x]^2\right)^{1/2}\right\} for\; x > L/2 \quad (6)$$

of the two ranges labeled with (5) and (6) for independent variable 'x' above discussed cylindrical shape when a homogeneous matter distribution is considered. For $x >> L/2 > R$, the asymptotic expression for the gravitational force field of a point mass, i.e. $-x^{-2}$, is recovered.

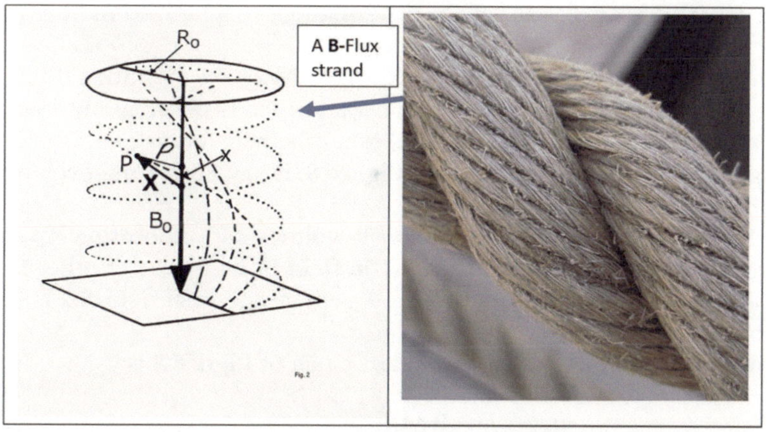

Figure 8.
Idealization of a structure of self-sustained ***B****-field MHD element and environment.*

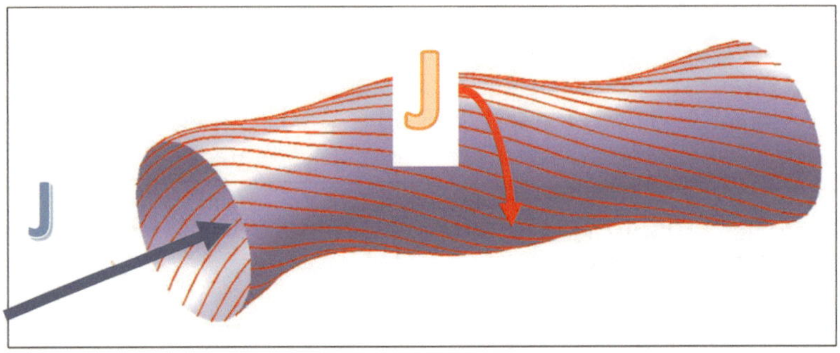

Figure 9.
The matter-magnetized in equilibrium flux tube idealization of a part of the LISM.

The approximate force expression at a distance ρ from the axis 'near' cylinders center is given by the expression:

$$F_\rho = -\nabla\phi)_\rho =$$
$$-\tfrac{1}{2}\, GN\rho \;\text{ for } 0 \le \rho \le R,$$
$$-\tfrac{1}{2}\, GNR^2\rho^{-1} \text{ for } R < \rho \le L/2, \tag{7}$$

In this case, the *Gaussian theorem* is used to obtain the above-approximated solution that can be used reasonably well in the range $0 \le \rho \le L/2$ for $R < L/2$ and $-L/4 \le x \le L/4$. The solution will be exact for any location in space, *although unrealistic*, for the self-gravitating field of the mass of the structure in the simplifying case of an infinite homogeneous cylinder of the medium.

Depending on the orientation of the magnetic field of the structures, if the conditions are right in an encounter of two magnetized tubes through their contact in their basis/tops extremes, the annihilation of the magnetic field is bound to occur (flux-tubes with opposed helicity/axial-currents) and the unbalanced attractive gravitational force will produce the matter agglutination which may be the cause of the so-called explosive formation of stars, i.e. cradle of the birth of stars as observed in astrophysics, see, e.g. [20, 21].

3.3 On the constitutive property of these permanent magnetized matter structures

Here we present our understanding on the conditions we identified through a variety of studies of low beta MHD from the low corona to observations in situ at 1 AU. This simple realization, our model, is a [22] flux tube with a twist, i.e. a flux rope (FR) free of magnetic stresses, see left panel in **Figure 8**. It has a 3-D time-evolving solution [23].

Figure 8, left shows a mass fraction Δm_P in volume ΔV_P at location $\mathbf{X_P} = \rho\, \mathbf{e_\rho} + x\, \mathbf{e_x}$ from the center of an FR with **axial magnetic field** B_0 and also *five* other **B**-field lines wrapping around the FR axis. The distance R_0 where the axial field changes sign is indicated on the top.

The analytical solution represented in left panel of **Figure 8** is

$$\mathbf{B} = B_0\, (r_0/r)^2\, \left[H\, J_1(a\, r)\, \mathbf{e_\varphi} + J_0(a\, r)\, \mathbf{e_x}\right], \tag{8}$$

Assuming homogeneous mass distribution; self-gravitational force

$$\mathbf{F}(\mathrm{r}, \varphi, \mathrm{x}=0) = -2\pi \mathrm{G}\, r\, \mathbf{e_r}\ (\mathrm{cyl}); \text{and for } \mathrm{r} << \mathrm{R}\ (\text{torus, not so everywhere in space}) \tag{9}$$

with the magnetic field expressed in polar coordinates, consistent with a right coordinate system, and for historical reasons, e_x is employed instead of e_z [24].

Then distortion in current/field(s) will happen, and magnetic stresses appear, as it is shown in [25], and there is *'no more a magnetic stress free MHD solution'*:

The **J** × **B** 'stresses', in addition is the magnetic force that equilibrates the region of interface between the local interstellar medium (LISM) and the sheath of the solar plasma (SP), as observed with Voyager 1.

Here we consider extremely short time scales, i.e. from a quarter of a day, to even year(s) when we think of the possible Hundreds of millions of years of existence of the molecular interstellar cloud which V1 and V2 explore in the very LISM as well as conditions at heliopause (HP) and the very LISM interface. At the start of the LISM (the local 'molecular cloud') [26] noticed the occurrence of a plasma depletion layer (PDL) in the LISM. The presence of such kind of PDL is a well-established feature at sheath—magnetosphere studies of Earth and other planets, see, e.g. Caan, McPherron, Russell, [27], Singh, [28].

An evaluation of **J** × **B** (magnetic stresses) equilibration of gravitation and pressure forces shows estimate studies to which we refer here considering:

1. pressure of gas (P_{gas}) at heliopause (hp) with solar origin ($3{\cdot}10^{-12}$ dyn)

2. self-gravitational Force at Helio-Sheath-LISM interface at *'hp'* (~10 orders less intense than above distortion)

3. self-organization as a 3-D Langmuir lattice (see, e.g. [29])

4. also a negligible ionization source by trapped cosmic ray isotropic population in the LISM has been measured (frequency of 10^{-17} $\sec^{-1}$ in [30]).

It is noted that the self-gravitational force, being attractive, would try to pull particles frozen in **B**-field together. This effect will be counteracted by the magnetic stresses to reduce the magnetic tension trying to keep a close equilibrium, i.e.,

$$\mathbf{J} \times \mathbf{B} + \mathbf{F}_g = 0 \tag{10}$$

A superconducting nature of the structure with a permeability much smaller than in vacuum is helpful to support larger amounts of mass in interstellar molecular clouds having a superconducting MHD nature.

Hence, to equilibrate the gas pressure (P_{gas}) in heliosheath implies same pressure from the magnetic field force, **J** × **B** in the LISM at hp interface and this requires a magnetic permeability of about (½ ± ¼) μ_0, detailed in the Appendix B.

This $\boldsymbol{\mu}$ would allow equaling magnetic force to the gravitational push of the Sun-system pressure (with the achievement of equilibrium) as it is required from observational studies, e.g. [19], see also [31].

An interesting theoretical perspective of the interaction has been attempted with quite a thorough mathematical approach in Usmanov et al. [32]. Usmanov et al. work considers the role of turbulence transport. Turbulence in MHD shock interfaces is assumed to be responsible for the acceleration of particles. The plasma pressure contribution in the sheath—LISM interface has its main contribution from these accelerated particles, e.g. in [19], and [31], better known in the literature as anomalous cosmic rays present in the heliosphere sheath region as discussed, e.g. in these two works.

We consider studies in which conditions show key similarities to the ones we just identified in the LISM: Long life duration of the structure, supportive of understanding that we are in the presence of self-sustained strong magnetic fields in a medium in which we took the initiative to investigate constitutive properties [18]. In a much longer line of work by several authors, it has been possible to describe MHD evolutionary behavior analytically of low beta MHD structure's basic properties, and there has been active research since the earlier 1990, see, e.g. [33–36].

With a full set of high-resolution data having a very low uncertainty in the instrumentation of the SC Wind, we were able to find/go beyond that and introduce ourselves to the constitutive manifestations of the low beta long-living MHD steady structure(s):

The magnetized MHD plasma stability was evaluated in a case study where it was identified as quite long duration and stable for an evolution in time and space in which the size changed by a factor 200 and consistent with preservation of its magnetic flux content suggesting isotropic behavior. A needed condition for the stability of the constitutive material and consistent with the *'no'* measurable presence of any process of diffusion of the magnetic field lines, or loss of it.

Of the structure of the strongly magnetized plasma medium, we can get information with the help of the observations of the Wind SWE instrument, as illustrated in **Figure 10** panels, left for the protons, from [37], and right panel for the electrons from Nieves-Chinchilla and Figueroa-Viñas [38]. Here, we have to focus on the narrow distributions, the left panel proton is a pure random distribution, but on the right panel electron distribution, we discuss solely the narrow part of the distribution. When doing that, we notice that **Figure 10** left and right panels for different strong **B**-field intervals like the one discussed in [18] makes plausible the here proposed 3-D Langmuir conditions of the medium when we observe that matter turns the **B**-field structure elastic, and with matter frozen independently to it, will show displacements consistent with the Hook's law of the oscillator, and statistically random (i.e., wholly

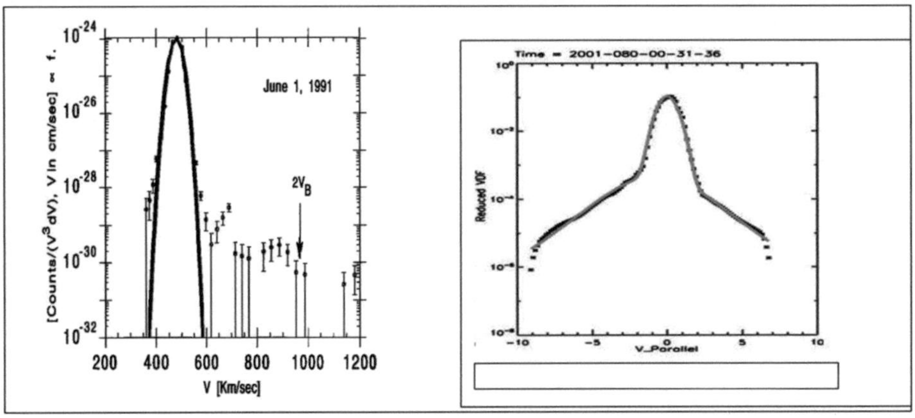

Figure 10.
'p' simpler and more complex 'e' density's structure illustrated by their Maxwellians.

independent, see, e.g. [39]). This behavior is well described by Langmuir (see, e.g. in Langmuir's adsorption theory of an *'ideal lattice gas,'* presented in a textbook on statistical mechanics by [29], and also [40]). This is understood here as a 3-D Langmuir amorphous crystal structure. And as we can simply show by partition energy under thermodynamic equilibrium **Figure 10** constitutes a strong support to the view as it follows:

The 'frozen matter' (electrons and protons here identified) for equal elastic strength will show a random distribution of the oscillating masses (see, e.g. [41]) with a mean amplitude which depends on the mass, for the same elasticity constant '*K*.' For the evaluation of the validity of the assumption, we analyze the proton and electron (narrow e-)distribution function shown in left and right panels of **Figure 10**. In this focused analysis of the distributions we notice first the pure random (vibration) motion of the frozen protons observed in the left panel of **Figure 10**. In the case of the electrons, we distinguish two clear regions, a narrow and a wide distribution. Here we interpret that frozen electrons are the narrow 'Gaussian' part of its distribution, and have once more the type of distribution in velocity consistent with random oscillation motion for frozen electrons and the same elasticity constant '*K*' as it corresponds to the crystalline amorphous medium assumed, see the detailed analysis in Appendix A.

Figure 10, right panel shows that when focused on the broad part of the *e-distribution,* if in equilibrium, there it gives information on temperature '*T*' from the *e-gas* of the structure (see, e.g. [42]). In our interpretation, the distribution of electrons, in **Figure 10**, right panel also has contributions from *e-lattice*, and *e-gas, as well as e-current aligned* ($\mathbf{J_e}$).

Here we are in the presence of an example beyond the notion of a frozen matter system in which far more complex features are shown for electrons than for ions and, in that way, giving room to further the investigation on the constitutive nature of this very stable matter as it is manifested in the presence of matter that through thousands of millions of years refused to agglutinate by the presence of gravitational laws of universal attraction between material objects.

The wider Gaussian (more commonly known as a Maxwellian in plasma astro physics) illustrates a different behavior for a subset of the electrons in the structure that, since the start of the study of the solar wind, has been a challenge to the interpretation of the worker in the field. In our case, we see that population as the dilute gas that pervades the medium, in a solid would be interpreted as the upper band that defines the metallic nature of the solid in consideration. We take that view which is well supported in the more detailed analyses we did in [18] and we proceed to detail below. There we expand on [43] pioneer description and understanding of the matter presence in magnetic clouds, *'strongly magnetized matter.'* (Section 3.4 further discusses the understanding achieved elsewhere on the central subject of this chapter: the MHD medium studied by Voyager SC in the very LISM.)

3.4 Properties and peculiar characteristics of the dilute e-gas of strong magnetized matter

We can further our observational understanding by doing a deeper analysis of the electron's distribution shown in **Figure 10**. And **Figure 11** reveals us more about this state of matter.

Figure 11 illustrates for a steady matter MHD case with intense/dominating **B**-field a set of 8 'skymaps' showing an angular map of the electron's distribution. In addition,

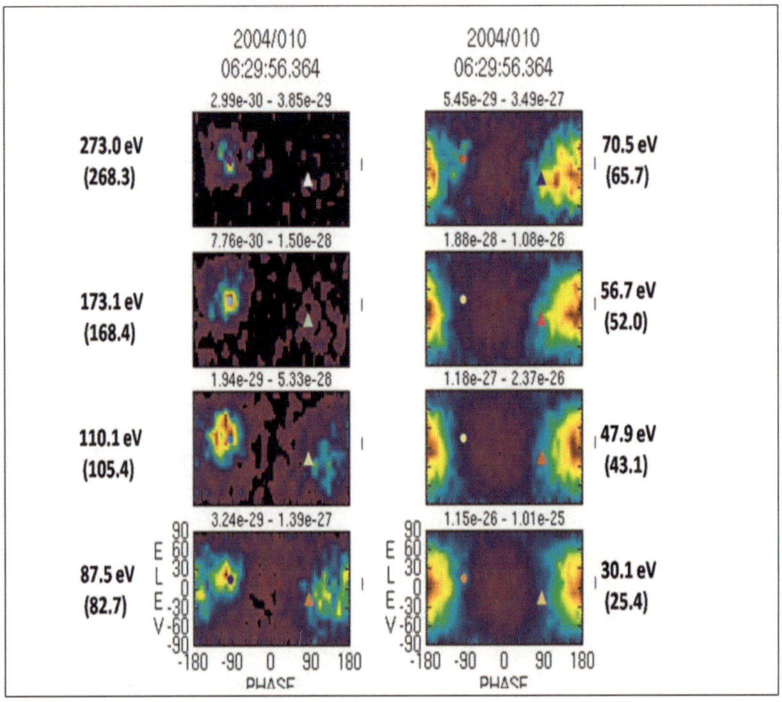

Figure 11.
Parts of e-vel distribution function F(v,θ,φ), *including Strahl aligned to* ***B****-field.*

point and arrow indicate in the plane the orientation of the **B**-field at the time interval considered, see Figure 1 in Figueroa Viñas et al. [44], see also [45].

The *'e' velocity* distribution function (VDF) in spherical coordinates is represented in **Figure 11** by a function f(v,θ, ϕ), where the independent variables are the velocity, polar angle (elevation in the figure), and azimuthal angle (Phase label, bottom, in the figure) respectively. **Figure 11** shows a subset not contiguous of the 30 energy steps of one VDF measurement, see instrument description in Ogilvie et al. [46]

The 2-D **Figure 11** choice of the longitude scale (from -180 to 180°) in each panel produces a split view of the selective one-dimensional distribution for MC (This distribution, along the **B**-field, is shown in **Figure 10**, right panel for the electrons (see ref. [38]). The location of the direction **B**-field is shown. The different panels below show from bottom up, on the right and continued on the left panels the intensity variation (from lowest values in lighter dark with increasing electron intensity through green, then yellow reaching the maximum intensity with red and most intense electron presence in dark red at the center of the distribution). Additionally, the plots show the orientation of the **B**-field direction through all panels (see arrowhead maintains the direction (θ = −30°, ϕ = 90°), and base of it seen in the right panes at (θ = 30°, ϕ = −90°), and indicates the current flowing in the direction opposite to the **B**-field intensity at energies of about more than 70 keV (all panels where the **B**-field electron aligned current J_e appears and in this case study extending beyond the upper energy interval shown of 268.3 keV.

Figure 12 shows that the *electrons* data are *'3s averages,'* also from the SWE instrument, but an instrument part built in GSFC (all in SC Wind) [47]. Fitting parameter

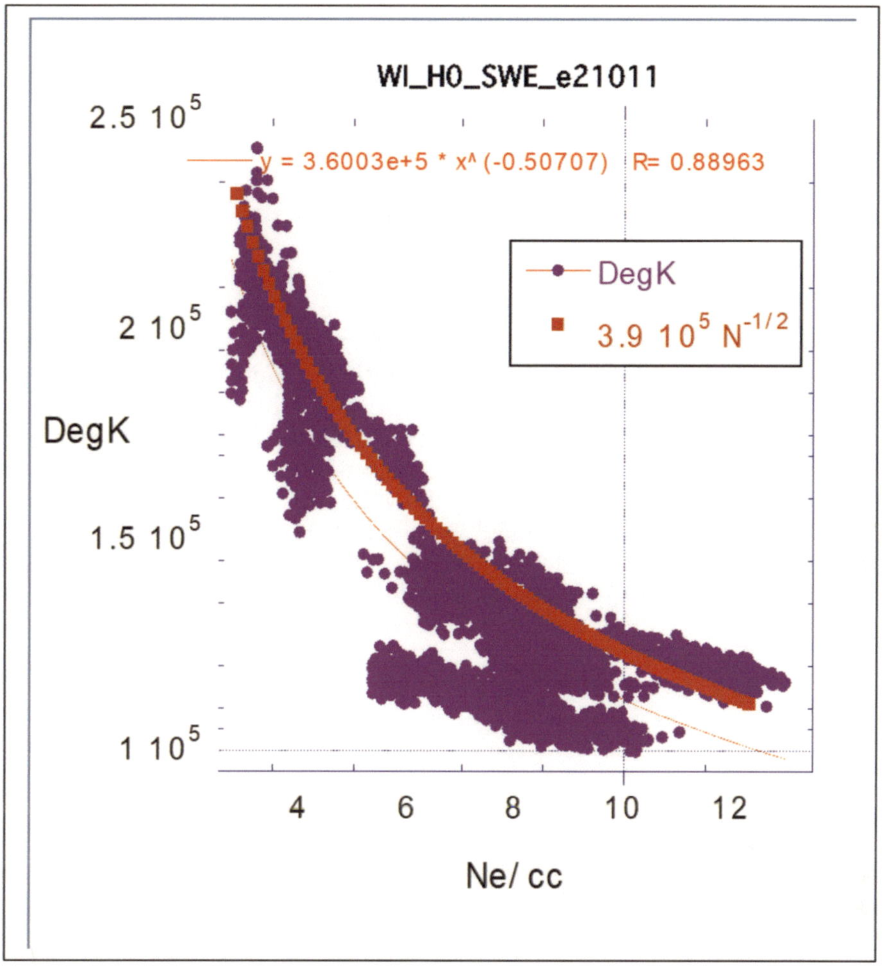

Figure 12.
*An example of anomalous gas properties in the SW for a **B**-field MHD medium.*

$\gamma = ½$ is the overall value of the polytropic exponent index on a strongly **B**-field dominated structure (MC interval with date 6/2/1998 observed) which passes Wind in its motion away from the Sun.

The γ for the FR region observed with Wind (see enclosed in black values on the right) would indeed show a smaller value for the relationship between temperature '*T*' of electrons and density of the plasma ($\gamma \sim 0.7$).

From a case study, *see* [18], obtained values of *T*, v_s, μ, $\mathbf{J}_{free}$. *And also:*

$$\mathrm{T} = 1.24 \cdot 10^5\ \mathrm{K} > 1/2\, m_e < (\boldsymbol{V_e} - < \boldsymbol{V_e} >)^2 > \text{which is proved right} \tag{11}$$

- $v_s(T) = 0.95{\cdot}10^5\ ms^{-1}$, *structure's acoustic speed, from e-gas relation* $v_s = [\gamma_a kBT/m_e]^{1/2}$
- $\mu(T)/\mu_0 = 0.5{\cdot}10^{-3}$, *from relation:* $\alpha = \mu\ \mathbf{H}{\cdot}d\mathbf{m}\ /\ PdV$ *and value of* $v_s(T)$
- $J_{free} = 5{\cdot}10^{-9}\ Am^{-2}$, *free current density, from* $\boldsymbol{J}_{free} = Ha/\mu\ \mathbf{B}$ *where the flux-rope **B**-field evolution is*

$$\mathbf{B} = \mathrm{B}_0\,(\mathrm{t}_0/\mathrm{t})^2\,\left[\mathrm{H}\,\mathrm{J}_1(\mathrm{a\,r})\,\mathbf{e}_\varphi + \mathrm{J}_0(\mathrm{a\,r})\,\mathbf{e}_\mathbf{x}\right], \tag{12}$$

own overall polytropic behavior on June 2, 1998, as it is shown in **Figure 12**.

The characteristics of the gas being of so considered anomalous kind, as the graph in **Figure 13** bottom panel shows. The top panel in **Figure 13** shows the usual ideal gas dependence of T (temperature) on ρ (density) for an adiabatic branch. It corresponds to the entropy conserving adiabatic process with polytropic exponent $\Upsilon = 5/3$, see, e.g. [48].

The actual observations summarized in **Figure 14** are based on a direct observation of the effect inside strongly magnetized matter in the form of a MC, already introduced earlier, suddenly ejected from the Sun and its interpretation based on data shown below in **Figure 14**. Our complete illustration in **Figure 14** of the **B**-field and bulk plasma conditions in an *electron* 3 s time series of a duration of a little more than two minutes of the properties present in this ideal **B**-field dominated matter includes from the top to bottom a sequence of eight panels ordered 1, 2, 3, 4, 5, 6, 7, and 8. In the present description: (a) panels 1 and 2 show the estimated convection velocity, in (1) the bulk velocity magnitude, named V_B, and in (2) on the left labels the longitude, and on the right the elevation of $\mathbf{V}_B$ in cthe orthogonal coordinate system GSE (e.g. [49]); (b) panels 3, 4, and 5 show the vector **B**-field, where panel 3

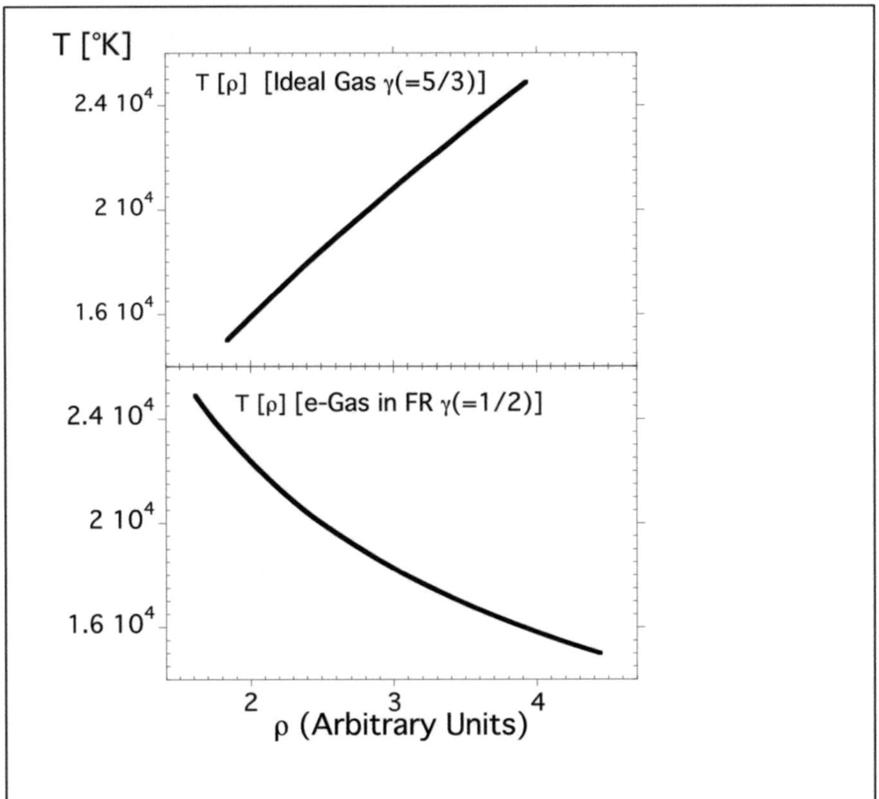

Figure 13.
Normal (top) vs. anomalous (bottom) gas.

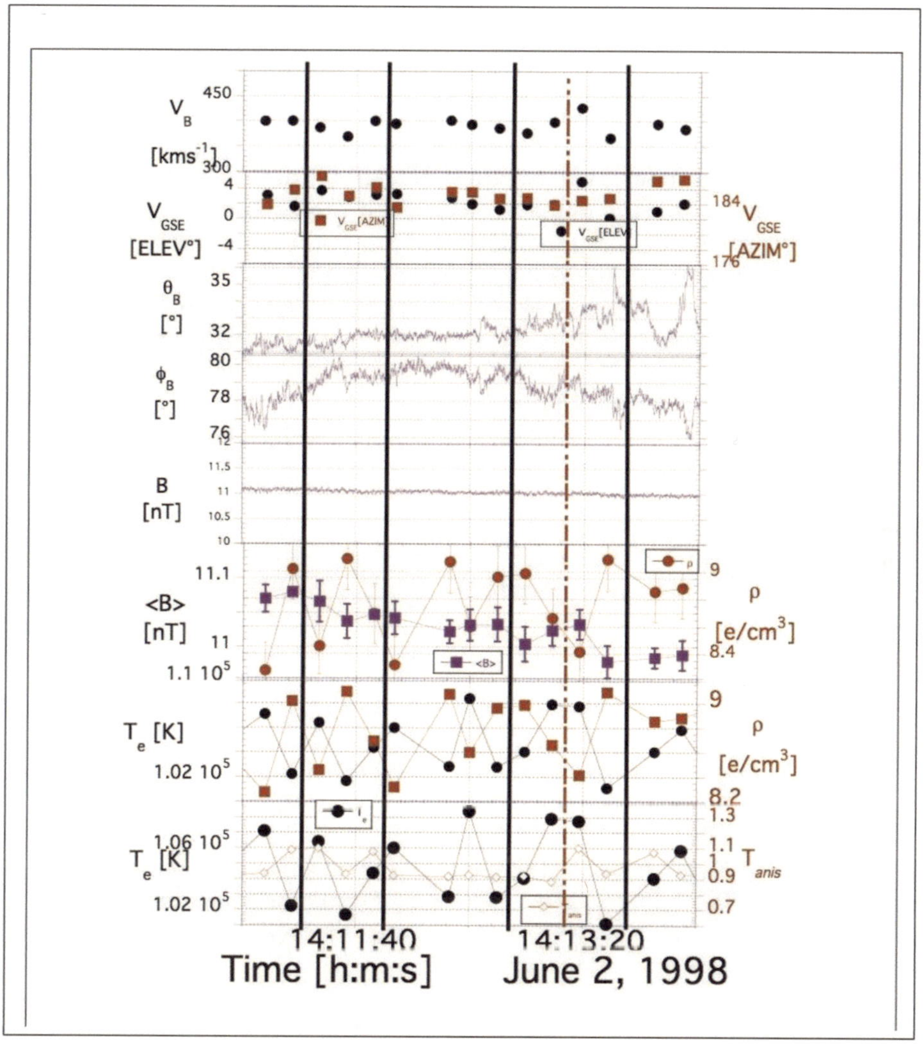

Figure 14.
Illustration of key constitutive properties of the MHD-dominated structure.

illustrates the steadiness of the intensity of the magnetic field, and the direction minimal changes are emphasized in panels 4 for elevation and 5 for longitude. Meanwhile, (c) panels 6–8 show the features of the medium in which panel 6 on the right shows the coherent variations showing anti-correlation between the 3 s mean **B**-field magnitude (on the left) and the electron density (on the right). Panel 7 on the right shows the clear anti-correlation between the electron T_e (left label) and density (right label); These panels 6 and 7 are at the core of the interpretation of a diamagnetic medium, consistent with past identification of this MHD constitutive property at interplanetary magnetic holes (see [50, 51]). Finally, (d) the bottom panel 8 checks the thermodynamic equilibrium condition showing negligible anisotropy in the internal electrons energy (right side of panel 8) while substantial changes, also shown in panel 7, (also illustrated in the left side of panel 8) take place indicating the diamagnetic nature of the medium, i.e. thermal equilibrium holds and further entropy conservation is well-established supporting the conservation of the energy of this matter–magnetic field structure. Further, the Figure illustrates the ends of the described coherence effects discussed at the passing of the time/location of the observation suggesting the presence of domains in the structure that would interact solely through their vibrational contact keeping the observed gas (wide part of the electrons distribution, **Figure 10** right panel, typical characteristic of the SW (rooted in the presence of a dilute gas in each domain), which constitutes a key element to understand the thermal short equilibration times in the low solar corona in our consistent interpretation of the here discussed medium, see, e.g. [52].

Figure 14 coherence of the diamagnetic coupling between plasma and **B**-field suggests again a presence in electron dilute gas of two works, which we discuss in the next paragraphs.

The presence of an anomalous polytropic index is the result of the two works done by the electron gas in the domains of the magnetized plasma structure (see [18, 52, 53]):

- ideal gas work only:

$$V^{\Upsilon} = Constant \qquad with\ \Upsilon = c_p/c_v = 1 + R/c_v = 5/3 \tag{13}$$

$$P\,V = N\,R\,T\ (equation\ of\ state)\ If\ N\ constant, then \tag{14}$$

$$T/\rho^{\gamma-1} = K \rightarrow with\ constant\ K = constant/\left(N^{\Upsilon}R\right) \tag{15}$$

Ideal gas and magnetic work: $\boldsymbol{\Upsilon} = 1 + R/c_v(1-\alpha)$ *where* α = *[magnetization work/gas work]*.

Then it can be shown the relationship between *e-gas* and magnetic work that enables anomalous γ ($\gamma_a = ½$) at position *P* in the FR in **Figure 8**. ... for a noninteracting *e-gas* with internal energy $dU = c_V\,dT$, and its equation of state $PV = RT$ (for 1 mole, and *R* being the universal gas constant) the relationship for the reversible adiabatic exponent γ_a is obtained from

$$c_V\,dT + R\,(1\text{–}\alpha)\,T\,dV/V = 0. \tag{16}$$

and the condition of point-like *e-gas* with three degrees of freedom, where $c_V/R = 3/2$, then $\alpha = 7/4$. (In Eqs. (13)–(16), *V* is the internal *energy* velocity of the gas and *not* the bulk velocity (V_B).)

4. Discussion; on the presence of large amounts of matter in gaseous state in galaxies

When we consider that in situ observation allows us to infer reasons for the stability of strongly magnetized matter as it is being explored by the in situ observations of V1 and V2, we may consider other presences of so-called molecular clouds and make educated argumentation for the lack of formation of small and large bodies on the argumentations of matter frozen to strong magnetic fields, which depending on relation of particle to magnetic field pressure we observe that condition well satisfied in the very LISM. The LISM is a structure that, for its characteristics, also seems not to collapse due to its own gravitation. Consequently, as we do here, it is possible to consider that the magnetic field's presence may be related to large amounts of matter in our galaxy, which stays in a gaseous form to be understood. In that respect, **Figure 15** illustrates a law of velocity of matter as a function of the distance of its center, which is common to many galaxies including ours. Distribution of velocities which resulted in easier capture from its outside.

Then, as we know, the result suggests the possibility that the common state of matter in the galaxies is of the dilute kind like the Voyager SC identified in our own local interstellar magnetized matter MHD state. We could speculate that perhaps in the neighborhood of the Sun, we could be in the presence of a formation of dilute matter old in nature, e.g. that in its origin constituted a proto-nebula as the one discussed in [55], a simple proto-nebula constituted of plasma under magnetohydrodynamic conditions. The presence of this kind of structure as a galaxy has indeed been observed, as illustrated in **Figure 16**.

Figure 15 shows, at scale, the galaxy view in the projected plane of the sky. Similar rotation velocities with the distance from the center have been identified for many galaxies, including the via Apia, our home galaxy. Notice the indicated departure with distance from the expected gravitational law when correlated with the matter's mass.

Here, it is possible to invoke that a proto-nebula as the one envisioned in [55] due to instabilities related to the cooling by radiation may have produced the instabilities responsible for partial collapse due to self-gravitational forces which generated the formation of stars, i.e. the evolution of the structure into a galaxy of the Hoag kind (Hoag's object).

When we make the standard assumption of considering the mass distribution in the ring galaxy (**Figure 16**) to be homogeneous, and we further consider that its larger amount happens to be in an MHD state, it is straightforward to write the analytical solution for the medium, where it should be only included the gaseous part of the mass and its attachment to the magnetic field in the type of magnetohydrodynamic nature we described for the proto-nebula in [55]. There the magnetic force solution is given by

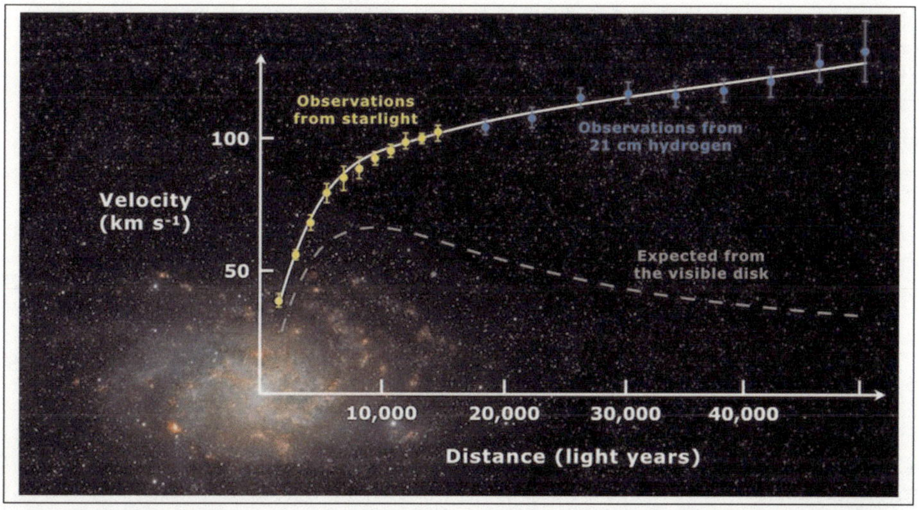

Figure 15.
Velocity of rotation in spiral galaxy messier Triangulum, see, e.g. [54].

Figure 16.
The Hoag object is a ring-shaped galaxy.

$$\mathbf{B} = B_0 \left[J_0(A(\rho,\phi))\ \mathbf{e_x} + H\, J_1(A(\rho,\phi))\ \mathbf{e_\phi}\right], \quad (17)$$

with argument

$$A(\rho,\phi) = a(\rho)\ [1 + \rho/R_{cFR}\ (cos\,(\phi) - |\,sin\,(\phi)|)] \quad (18)$$

would be the time stationary limit of the solution presented earlier [18, 25] in a different context, where $a(\rho) = A\ \rho$, and $A = j_0/R_{FRcore}$, $J_{0,1}$ are the well-known orthogonal, grade 0 and 1, cylindrical Bessel functions of first kind[9], and the value of $\rho = R_{FRcore}$ defines the radius of *'the circular cross section'* of an approximated cylinder-section of the torus at the location where the axial magnetic field has its first node, is identically zero, *i.e.*, $J_0(j_0) = 0$. In this case study, there is no volume (*Vol*) change of the matter-magnetized structure with time (**Figure 17**).

Next we drop subindices for R_{cFR}, and instead, we use R. The summatory of all 'torus' elements and gravitational forces acting on each infinitesimal element 'δM_{Torus} 'at the torus' locus is oriented to the center of the torus, defined as

$$\mathbf{F}_g\ (\delta M_{Torus}) = -G\ M_{Torus}\ \delta M_{Torus}\ \mathbf{R}(\delta M_{Torus})/|\mathbf{R}|^3 \quad (19)$$

for circular symmetry of the simple case of homogeneously distributed matter in a magnetized field (defined in Eq. (17)). Also, any mass particle m located at a distance from the center of the torus $|\ \mathbf{r} - \mathbf{R}|$ is bound to feel such gravitational pull,

$$\mathbf{F}_g\ (m(\mathbf{r}-\mathbf{R})) = -G\ M_{Torus}\ m\ (\mathbf{r}-\mathbf{R})/|\mathbf{r}-\mathbf{R}|^3 \quad (20)$$

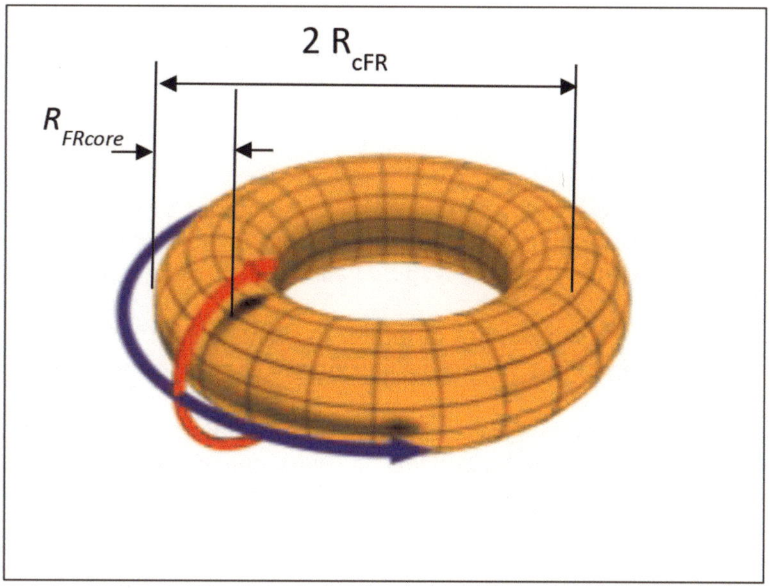

Figure 17.
*A torus-shaped solution to **B**- and self-gravitational fields stable equilibrium.*

Once with the expression for the magnetic field and the current, it is easy to find the equilibrium condition for

$$\sum \mathbf{F} = \mathbf{0} \tag{21}$$

where we here solely consider magnetic and gravitational forces, i.e.

$$\int_{M} \mathbf{F}_g \left(\mathrm{dM}_{\mathrm{Torus}}\right) \mathrm{dM} + \int_{J} \mathrm{d}\mathbf{J}_C \, \mathrm{X} \, \mathbf{B} = \mathbf{0} \tag{22}$$

and, considering the solution of Eq. 17 in [25] for a truncated torus, we proceed to the generalization of the simpler case of a whole torus developed here, which for us, corresponds to the simplified equilibrium expression:

$$G\,(M_{Torus})^2 + (25\,\mu_0/4)\left[\Phi_\phi{}^2\right] = 0. \tag{23}$$

The mass of the torus (M_{Torus}) and the poloidal magnetic flux (Φ_ϕ) forces in equilibrium are simply related, as shown in Eq. (23). In this way, we obtain, for a simple limit, the condition for the matter to be attached to the 'protogalaxie magnetic field' generated by convective currents, which stabilize the matter in a cold *frozen matter* condition. Preliminary studies [18, 51] suggest that this, '*stable structure*,' will possess specific thermodynamic properties characteristic of a diamagnetic environment. A possible explosive disruption of this stable condition is addressed below.

4.1 From the Hoag's object to spiral galaxies

A process of destabilization of the structure could be caused by the interaction of two proto nebulas of the kind described above, producing changes to the morphology of the simple formulations presented. In such a collision scenario of two rings Hoag's nebula or proto-nebula, there would be gravitational cause for strong dislocation of the toro generating reconnections between magnetic domains with opposing polarity and causing a spreading in regions presenting a collapse of the structure MHD in this way freeing the matter in multiple locations of the fresh forming spiral galaxy in an explosive process of formation of stars with the same birthmark in their matter ratio of elements constitutive inside of the new nebula/galaxy.

This origin, with still the presence of a majority of matter in a dilute state frozen to the magnetic field, clearly gives a straightforward explanation of the no Keplerian distribution of velocities in the known spiral galaxies, including our Via Apia, the home of our solar planetary system.

In the actual galaxies, the amount of matter available would be far from being exhausted (see Star Formation Sputtering Out Across the Universe by Space.com Staff , November 07, 2012).

The majority of the known galaxies would still be holding their matter in a stable MHD state of matter frozen to the **B**-field. The consequence of this would be the presence of a fragmentary toro-shaped dilute cold matter (<10,000 K), which occasionally would form new stars in renewed disruptive processes, see, e.g. [20], generating the so-called 'cradles of the new stars.'

5. Conclusions: about our learning with the in situ observation using Voyager SC

In this chapter, it is probed a variety of key properties of the local to solar system interstellar molecular cloud and its millions-of-year presence of a very strong magnetic field. The proof was achieved by the in situ measurements of a variety of its properties for a whole set of data recollected by several instruments in the mission Voyager (V1 and V2 SC) as we reported: Magnetometer, Radio Instrument, Plasma Instrument, low and high energy particle instrument of the Voyager SC. In this way, the Voyager SC can identify the presence of a medium that has very long-lasting permanent magnet characteristics and an origin foreign to the solar magnetized plasma, which has its source in the solar corona.

Further, it is presented the consistency of the interpretation of the medium being matter in a strong **B**-field with a 3 D Langmuir amorphous crystal gas. The role of the frozen matter, the dilute electrons, and the presence of B-field convected currents is the source of a very stable magnetic field, as we know from some simple modeling illustrated in Section 4.

Section 3 discussed the identification of a consistent constitutive permeability property differing from the permeability of the vacuum smaller by a factor ½ appears to be the case considering its consistency by observations by V1 and V2. (See V2 preliminary estimate in Appendix B.)

A simple estimate is made of the mass of the local molecular clouds in which the Sun, Alpha Centauri, and a few other stars found in its neighborhood of the sketched molecular cloud complex, which also approximately co-move with it around the galaxy center.

Finally, in our discussion (Section 4), we extrapolate the interpreted properties of the medium to a possible origin of our home galaxy and other structures observed in astrophysics.

Among key subjects in need of further study are: (a) energy/nature arguments on the reasons for nonradial orientation of the magnetic **B**-field at heliopause in the heliosphere sheath and very LISM. A matter which is still in need of understanding. (b) Stronger validation of assumptions made in this work which so far offer consistent support of the nature of the permanent magnetized state of matter here introduced to a wider range of readers until now mostly limited to members of the geophysical and astrophysical communities. (c) The role and range of cosmic rays' possible modification of the properties of the MHD structure.

Acknowledgements

The collaboration with all team members, also the consideration by the current instrument manager Adam Szabo for my work is warmly thanked. Continuous NASA NNG15CR65C grant support to Voyager SC magnetometer enabled and supported the expenses of this book contribution. I acknowledge the benefit of my learning and the opportunity to study due to my father's love, the late Carlos David Grünfeld Berdichevsky.

A. Appendix

Essential differences are present in each panel of **Figure 10**, there are both plots, which are simply related to the elaboration pursued in their generation. A panel of the left for the protons gives just the proton flux observed in a steady (undisturbed) strong **B**-field interval as a function of a very well-collimated radial velocity from the Sun of the proton population selected through a process of triple coincidence that includes detectors independent evaluation of the start time each proton enters, followed of the time of impact in the stopping location, and charge, mass, and energy by a strong magnetic field sweep that by deflection of the particle identifies the ion's unique properties of interest, see, e.g. [56]. Further, the plot evaluates the intensity of the proton's flux from the Sun as a physical invariant property of the interval of observation which is extremely small and consistent with the well-known high vacuum of the interplanetary medium near the distance Ulysses is found at the time of observation in its initial trajectory from the Earth to Jupiter, as indicated in the work of these observations, i.e. [37]. While in the case of the very detailed study of the electron distribution in Nieves-Chinchilla and Figueroa Viñas [38] with the SWE instrument in Wind SC

$$< \text{Energy}\left(\text{electron} \sim \tfrac{1}{2}\, m_e \left(0.8\ \text{kms}^{-1}\right)^2\right) / \text{Energy}\left(\text{proton} \sim \tfrac{1}{2}\, m_p \left(21\ \text{kms}^{-1}\right)^2\right)^* \\ \text{mass(p)/mass(electron) is in the order of 1836.6.} \tag{24}$$

But still, we can do one more normalization related to the general radiative cooling of the strong magnetized matter, which increases with SC distance from the Sun, which for Ulysses is close to 3 AU from the Sun versus 1 AU for the location of Wind SC, and the time of the respective measurements. This is valid when we consider the radiative cooling of matter with time in the interplanetary space and relate the observations with Ulysses of the proton. This behavior is well understood for a frozen matter estate of nature which depending on its mass produces tighter oscillation for smaller mass; versus larger oscillation amplitude for larger masses. Henceforth, strong support for the constitutive nature of the ideal magnetized matter is identified through very delicate analysis using very clean SWE in Wind SC, and SWICS in Ulysses SC plasma instruments. Here we deal with far smaller uncertainties resolution than what is available with most instruments from past and planned missions.

Hence, the obtained ratio close to one is a quite strong argument favoring the 3-D Langmuir amorphous crystalline nature of the low beta ideal MHD matter structure.

B. Appendix

On the subject of units, we start with the relationship of equivalence for the pressure of

$$1\,\text{dyn/cm}^2 = 0.1\ \text{Pascal}, \tag{25}$$

and considering that 1 fPa = 10^{-15} Pa. We notice that [19] reached a value for the particle pressure pushing the surrounding LISM by the heliopause with a pressure intensity of $3.57 \pm 0.71{\cdot}10^{-13}$ Pa, and in its independent work, [31] proposed the observational value of 267 ± 55 fPa. This is the same that $2.67 \pm 0.55{\cdot}10^{-13}$ Pa at heliopause crossing by V1. In a way agreeing within less than two standard deviations the two studies. Here we take advantage of the consistency of both studies to assume a pressure onto the LISM from the heliosphere of 300 ± 50 fPa, which would correspond to an equilibration local interstellar magnetic field of 0.7 ± 0.1 nT, which is about a much higher value of the average magnetic field of the very LISM of about 0.4 nT and which can simply be explained by our understanding of a medium which due to an anomalous electrons gas has as a consequence through the presence of magnetization work coupling to the gas pressure a medium permeability μ close to ($1/2 \pm 1/4$) μ_0. *Equilibrium with* μ*!*

Burlaga et al. [47] show in V2 an '*expected*' **B**-field barrier at heliopause with a **B**-field intensity of approximately 0.45 nT on the sheath side of the Heliopause. On the LISM side, the **B**-field is $\sim$0.7 nT. This implies a push by the sheath of about 500 fPa onto the interstellar medium, while at '*hp*' interface, the LISM **B**-field stresses equilibrate that push for a similar small value for the constitutive permeability of the medium $\mu \sim 1/2\, \mu_0$ (see, e.g. [19]). This is consistent with our interpretation of the presence of magnetization work discussed in Section 3.4 with the help of more detailed studies from more advanced experimental quality plasma and equally good magnetic field data available from Ulysses and Wind SC.

Author details

Daniel Benjamín Berdichevsky[1,2]

1 IFIR at UNR-CONICET, Esmeralda y 27 de Febrero, Rosario, Sta Fe, República Argentina

2 Aleyon Technical Services, JV, LLC, USA

*Address all correspondence to: dbberdi@gmail.com

References

[1] Stone E. Voyager 1 has entered interstellar medium, news announcement at NASA headquarters, Washington, DC, USA. EOS. 2013;**94** (39):339. DOI: 10.1002/2013EO390003

[2] Behannon KW, Acuña MH, Burlaga LF, Lepping RP, Ness NF, Neubauer F-M. Magnetic field experiment for Voyager 1 and Voyager 2. Space Science Reviews. 1977;**21**:235-257. DOI: 10.1007/BF00211541

[3] Bridge HS, Belcher JW, Butler RJ, Lazarus AJ, Mavretic AM, Sullivan JD, et al. The plasma experiment on the 1977 Voyager Mission. Space Science Reviews. 1977;**21**:259-287

[4] Krimiges SM, Amstrong TP, Axford WI, Bostrom CO, Fan CY, Gloeckler G, et al. The low energy charged particle (LECP) experimenton the voyager spacecraft. Space Science Reviews. 1977; **21**(3):329-354. DOI: 10.1007/BF00211545

[5] Scarf FL, Gurnett DA. A plasma wave investigation for the Voyager mission. Space Science Reviews. 1977;**21**:289-308

[6] Stone EC, Vogt RE, McDonald RE, Teegarden FB, Trainor JH, Jokipii JR, et al. Cosmic ray investigation for the Voyager mission; energetic particles studies in the outer heliosphere and beyond. Space Science Reviews. 1977;**21**: 355-376

[7] Berdichevsky DB. Survey of Compressions Observed in the Heliosheath with the Spacecraft Voyager. Advances in Aerospace Science and Technology. 2021;**6**:43-66. DOI: 10.4236/aast.2021.61004

[8] Burlaga LF, Kurth WS, Gurnett DA, Berdichevsky DB, Jian LK, Ness NF, et al. Magnetic field and plasma density observations of a pressure front by voyager 1 during 2020 in the very local interstellar medium. The Astrophysical Journal. 10 April, 2021;**911**(61):9. DOI: 10.3847/1538-4357/abeb6a

[9] Holt EH, Haskell RE. Foundations of Plasma Dynamics. NY, NY: The MacMillan Co.; 1968

[10] Tidman DA, Krall NA. Shock Waves in Collisionless Plasmas. New York: Wiley; 1971. 166 pp

[11] Mostafavi P, Zank GP. The structure of shocks in the very local interstellar medium. Astrophysical Journal Letters. 2018;**854**:L15. DOI: 10.3847/2041-8213/aaab54

[12] Opher M, Loeb A, Drake J, Toth G. A small and round heliosphere suggested by magnetohydrodynamic modelling of pick-up ions. Nature Astronomy. 2020;**4**: 675-683

[13] Izmodenov V, Alexashov D, Myasnikov A. Direction of the interstellar H atom inflow in the heliosphere: Role of the interstellar magnetic field. Astronomy and Astrophysics. 2005;**437**:L35-L38

[14] Pogorelov NV, Heerikhuisen J, Zank GP. Probing heliospheric asymmetries with an MHD-kinetic model. Astrophysical Journal Letters. 2008;**675** (1):L41-L44. DOI: 10.1086/529547

[15] Jackson JD. Electrodinámica Clásica. In: Translated from English by Marsal Alexaindree, G. Madrid, Spain: Alhambra; 1966

[16] Alfvén HG. Existence of electromag netic-hydrodynamic waves. Nature. 1942;**150**:405

[17] Chew GF, Goldenberger ML, Low FE. The Boltzmann equation and the one-fluid hydromagnetic equations in the absence of particle collisions. Proceedings of The Royal Society of London. 1956;**236**(Series A):112-118

[18] Berdichevsky DB, Schefers K. On the thermodynamics and other constitutive properties of a class of strongly magnetized matter observed in astrophysics. The Astrophysical Journal. 2015;**805**(70):13. DOI: 10.1088/0004-637X/805/1/70

[19] Gloeckler GJ, Fisk LA. Energetic neutral hydrogen observations demonstrate that Voyager 1 is not observing the extraordinarily strong interstellar magnetic field. The Astrophysical Journal. 2016;**833**:290

[20] McKee M. Planetary science: Caught in the Act. Nature. 2013;**493**(7434):592-596

[21] Specktor B. Hubble image reveals the mighty baby stars that will probably destroy the 'Pillars of Creation'. LiveScience. 2020. Available from: https://www.livescience.com/pillars-of-creation-infrared-view.html

[22] Lundquist S. Magnetohydrostatic fields. Arkiv för Fysik. 1951;**2**:61

[23] Berdichevky DB, Lepping RP, Farrugia CJ. On geometric considerations of the evolution of magnetic flux ropes. Physical Review. 2003;**E67**:036405. DOI: 10.1103/PhysRevE.036405

[24] Lepping RP, Jones JA, Burlaga LF. Magnetic field structure of interplanetary magnetic clouds at 1 AU. Journal of Geophysical Research. 1990; **95**:11957-11965

[25] Berdichevsky DB. On fields and mass constraints for the uniform propagation of magnetic-flux-ropes undergoing isotropic expansion. Solar Physics. 2013; **284**:245-259. DOI: 10.1007/s11207-012-0176-5. Erratum. Solar Phys. 2019;**294**: 167. DOI: 10.1007/s11207-019-1557-9

[26] Gurnett DA, Kurth WS, Burlaga LF, Ness NF. In situ observations of interste llar plasma with Voyager 1. Science. 2013;**341**:1489

[27] Caan MN, McPherron LF, Russell CT. Characteristic of the Association between the interplanetary magnetic field and substorms. The Journal of Geophysical Research. 1982;**82**(ISEE issue A4):4837-4841, and other work on magnetopause in same ISEE issue

[28] Singh RN. The Earth's magnetopause and related phenomena. Indian Journal of Radio and Space Physics. 1987;**16**: 161-171

[29] Hill TL. An Introduction to Statistical Thermodynamics. Reading, Massachusetts, US: Addison-Wesley; 1960

[30] Cummings CH, Stone EC, Heikkila BC, Lal N, Webber WR, Óhannesson GJ, et al. Galactic cosmic rays in the local interstellar medium: Voyager 1 observations and model results. The Astrophysical Journal. 2016;**831**:18(21 pp). DOI: 10.3847/0004-637X/831/1/18

[31] Rankin JS, McComas DJ, Richardson JD, Schwadron NA. Heliosheath properties measured from a Voyager 2 to Voyager 1 transient. The Astrophysical Journal. 2019;**883**:11. DOI: 10.3847/1537-4357/ab3d9d

[32] Usmanov A, Goldstein ML, Matthaeus WH. A four fluid model of the solar wind/interstellar medium interaction with turbulence transport and pickup protons as separate fluids. The Astrophysical Journal. 2016;**820**

[33] Farrugia CJ, Osherovich VA, Burlaga LF. Magnetic flux rope versus the spheromak as models for interplanetary magnetic clouds. Journal of Geophysical Research. 1995;**100**:12293-12306

[34] Osherovich VA, Farrugia CJ, Burlaga LF, Lepping RP, Feinberg J, Stone RD. Journal of Geophysical Research. 1993; **98**:15331

[35] Osherovich VA, Fainberg J, Stone RG, MacDowall RJ, Berdichevsky DB. Self-similar evolution of interplanetary magnetic clouds and Ulysses measurements of the polytropic index inside the cloud. In: Wilson A editor. European Space Agency, Proceedings of the 31st ESLAB Symp., Ed at ESTEC. Noordwijk, The Netherlands: Correlated Phenomena at the Sun, in Heliosphere and in Geospace, ESA-SP 415. 1997. 171-175

[36] Osherovich VA, Fainberg J, Stone RG, FItzenreiter R, Viñas AF. Measurements of polytropic index in the January 10-11, 1997 magnetic cloud observed by wind. Geophysical Research Letters. 1998;**25**:3003-3006

[37] Ogilvie KW, Geiss J, Gloeckler G, Berdichevsky D, Wilken B. High velocity tails on the velocity distribution of solar wind ions. Journal of Geophysical Research. 1993;**98**:3605-3611

[38] Nieves-Chinchilla T, Figueroa-Viñas A. Solar wind electron distribution functions inside magnetic clouds. Journal of Geophysical Research. 2008;**113**: A02105. DOI: 10.1029/2007JA012703

[39] Kac M. Statistical Independence in Probability Analysis and Number Theory. The Carus Mathematical Monographs. Number 12. US: The Mathematical Association of America (MAA); 1959

[40] Langmuir I, Chemistry S. Nobel Prize Lecture, Dec. 14. 1932

[41] Goldstein H. Classical Mechanics. Cambridge, UK: Addison Wesley; 1956

[42] Sittler EC Jr, Burlaga LF. Electron temperatures within magnetic clouds between 2 and 4 AU: Voyager 2 observations. Journal of Geophysical Research. 1998;**103**:17447-17454

[43] Burlaga LF. Magnetic clouds and force-free fields with constant alpha. Journal of Geophysical Research. 1988; **93**:7217-7224

[44] Figueroa-Viñas A, Gurgiolo CA, Nieves-Chinchilla T, Goldstein ML. Solar Wind Halo Formation by the Scattering of the Strahl via Direct Cluster/PEACE Observations of the 3D Velocity Distribution Function, Contribution to Solar Physics NTRS NASA Meet2010. DOI: 20100026391

[45] Fitzenreiter RJK, Ogilvie W, Chornay DJ, Keller J. Observations of electron velocity distribution functions in the solar wind by the WIND spacecraft: High angular resolution strahl measurements. Geophysical Research Letters. 1998;**25**(3):249-252

[46] Ogilvie KW, Chornay DJ, Fitzenreiter RJ, Hunsaker F, Keller J, Lobell J, et al. Space Science Revies. 1995; **71**:55-77

[47] Burlaga LF, Ness NF, Berdichevsky DB, Jian LK, Park J, Szabo A. Magnetic field observations in the very local interstellar medium by voyagers 1 and 2. The Astrophysical Journal. 2022;**932**:59. DOI: 10.3847/1538-4357/ac658e

[48] Zemansky MW. Heat and Thermodynamics. NY, NY: McGraw-Hill Book Co.; 1957

[49] Mish WH, Green JL, Reph MG, Peredo M. ISTP science data systems and

products. Space Science Reviews. 1993; **71**:815-877

[50] Burlaga LF, Lemaire J. Interplanetary magnetic holes: Theory. Journal of Geophysical Research. 1978; **83**:5157-5160

[51] Lemaire J, Burlaga LF. Diamagnetic boundary layers: A kinetic theory. Astrophysics and Space Science. 1976;**49**: 303-325

[52] Berdichevsky DB, Rodriguez JM. Thermodynamic interpretation of Electron density and temperature description in the Solar Corona. Advances in Aerospace Science and Technology. 2022;**07**(03):146-174. DOI: 10.436/aast.2022.73010

[53] Berdichevsky DB, Gomez JMR, Veira LM, Lago AD. Validation of novel model for identification of thermal conditions in the low corona. Advances in Aerospace Science and Technology. 2022;**7**:52-84. Available from: https://www.scirp.org/journal/aast

[54] Rubin V, Thonnard N, Ford WK Jr. Rotational Properties of 21 Sc Galaxies With a Large Range of Luminosities and Radii from NGC 4605 (R=4kpc) to UGC 2885 (R=122kpc). The Astrophysical Journal. 1980;**238**:471-487. DOI: 10.1086/158003

[55] Berdichevsky DB. A simple proto nebula constituted of plasma under magneto hydrodynamics conditions. Aero Aero Open Access Journal. 2022;**6** (1):22-24. DOI: 10.15406/aaoaj.2022.06.00137, May 30

[56] Gloeckler G, Geiss J, Balsiger R, Bedim P, Cain JC, Fisher J, et al. The solar wind ion composition spectrometer. Astronomy and Astrophysics, Supplement Series. 1992; **92**:267-289

Chapter 6

Current Advances in Nanocrystalline Rare Earth Based Modern Permanent Magnet

Dipti Ranjan Sahu

Abstract

Modern permanent magnets are the essential materials for many industries and technologies. All strong permanent magnets that contain rare earth element have wide range of application. Different processing technique, material and manufacturing methods are used to produce different types of rare earth magnets. New developments and improvement in properties are proposed based on the addition of nanocrystalline materials to address the effectiveness of rare earth magnets which is beneficial for different application. This chapter reviews the fundamental aspect and progress of rare earth modern magnet along with the need of essential key factor for future development of cost-effective rear earth permanent magnets.

Keywords: magnets, rare earth, nanocrystalline, NdFeB, Sm-Co, SmFeN, sintering

1. Introduction

Permanent magnets are critical components for design of modern devices in many technological aspects. Modern permanent magnets made of rare earth magnetic materials play important role in clean energy sector and climate economy products such as electric vehicles, consumer electronics, wind turbines, military products, phones, refrigerators, weapons, missiles and jets [1–4]. Rare earth permanent magnets consist of some rare earth elements (Sm, Nd, Pr, Dy, Ce, etc.) form critical magnetic structures. Most of the research approaches have been dedicated towards synthesis of permanent rare earth materials, its characterization, modeling, and processing of materials [5–7]. Such materials have specific magnetic order parameter which define its physical properties for potential applications. These properties are extremely important for our daily life: exploiting the benefit of mentioned promising order parameters (coercivity, spontaneous magnetization, Curie temperature), it is possible to develop and design efficient technological devices for computer and office automation, household appliances, and magnetic data storage, position sensors, actuators, micromechanical system and medical application [4].

It is widely known that Sm-Co magnet was the first developed powerful rare-earth magnets and rapid advances was made on the improvement of the performance of Sm-Co magnets [8]. This was followed by the development of stronger and cheaper Nd-Fe-B rare earth magnets [4, 9]. In recent years, SmFeN magnets have gained

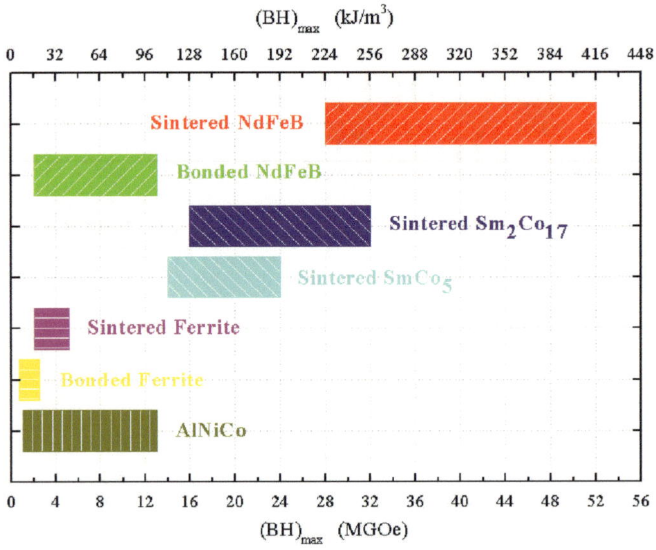

Figure 1.
The magnetization characteristics of some rare earth magnets (Ref: https://www.advancedmagnets.com/what-are-rare-earth-permanent-magnets-repms/.

attention due to their high-temperature stability and excellent magnetic properties [10]. SmFeN magnets are also known as third-generation permanent magnets and are considered to be an alternative to NdFeB magnets for certain applications.

Figure 1 presents the magnetization characteristics of the mostly used rare earth magnets (NdFeB and Sm-Co) with respect to non-rare-earth magnets. Nowadays Nd-Fe-B magnets received most attention in the market due to their cheap price and excellent room temperature magnetic properties. However, for high temperature and next generation application, high Coercivity is the prime requirement. The anisotropy field of the magnets can also be enhanced due to heavy rare earth magnets. The inclusion of heavy rare-earth magnets increases the cost of the magnets and reduces the remanence value. Sm-Co rare earth magnets have high curie temperature and high thermal stability. This magnet is the best candidate for high temperature applications. Similarly, Sm-Fe-N magnets which have higher anisotropy has low stability at high temperature which limits its practical application in high tech industries and energy transition technologies.

Therefore, continuous improvement of permanent magnet in terms of performance, efficiency and application is a technological challenge. The development of advanced permanent magnets is always a requirement which can provide high efficiency and reliability, low cost, outstanding magnetic properties and low maintenance for applications. This chapter reviews the fundamental aspect and progress of the rare earth modern magnet which highlight the essential key factor for future development of cost-effective permeant magnets.

2. Structure of the rare earth magnets

The three family of the rare earth magnets are (i) Sm-Co Magnets ($SmCo_5$, Sm_2Co_{17}), (ii) Nd FeB magnets (specially Nd-$Fe_{14}B$) and Sm-Fe-N magnets. These rare earth permanent magnet materials possess uniaxial crystal structure (tetragonal, hexagonal, or rhombohedral) and the magnetization along the crystallographic symmetry axis. Detail crystal structure are as follows,

2.1 Sm-Co magnets

Sm-Co magnet is a rare earth magnet consists of two basic elements samarium and cobalt. The phase composition and structure define the basic magnetic behavior of the magnets. The sintered 1:5 type SmCo magnets have hexagonal crystal structure or 2:17 type SmCo magnets have a rhombohedral crystal structure [11]. **Figure 2** presents the crystal structure of $SmCo_5$, $SmCo_7$ and Sm_2Co_{17}. The SmCo5 have the hexagonal $CaCu_5$-type crystal structure [12–15]. $SmCo_5$ structure belongs to the space group P6/mmm while Sm_2Co_{17} (P63/mmc) is Th_2Ni_{17}-type hexagonal or Th_2Zn_{17}-type rhombohedral ($R\bar{3}m$) structure. SmCo5 is a layer-by-layer configuration along the c-axis [16].

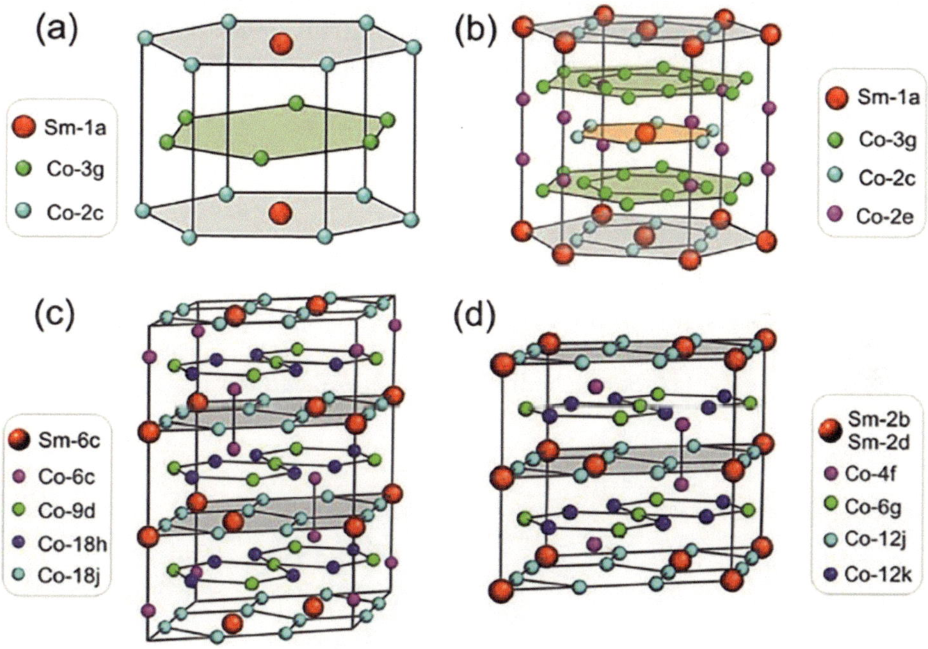

Figure 2.
Crystal structure of (a) $SmCo_5$ and (b) $SmCo_7$ (c) rhomohedral Sm_2Co_{17} (d) heaxagonal Sm_2Co_{17} [12–15].

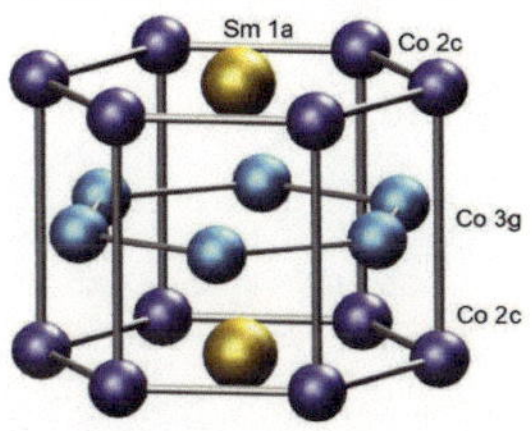

Figure 3.
Unit cell of the hexagonal $SmCo_5$ [17].

Typical unit cell of Sm-Co_5 is shown in **Figure 3**. Alternative Sm-Co layers and Co form the hexagonal structure. The lattice constants of stable bulk SmCo5 are a = 5.00 Å and c = 3.96 Å and of Sm_2Co_{17} are a = 8.36 Å and c = 8.15 Å [17].

The composition of distinct phases, their elemental profiles, and the process parameters optimization is necessary for the creation of perfect structure. The coherent crystal symmetry along with perfect phase separation is the key to achieve the technical realization of the unique magnets. The formation of such unique structure consisting of phases produce Sm-Co magnets with high Curie temperature for the high temperature operation.

2.2 Nd-Fe-B magnets

The rare earth element Neodymium along with Iron and boron alloy with $Nd_2Fe_{14}B$ have tetragonal structure permanent magnets. This tetragonal $Nd_2Fe_{14}B$ belongs to the space group $P4_2/mnm$. The lattice parameters values are a = 8.818(2) and c = 12.236(5) Å. This NdFeB shows large coercivities and energy product [18].

Figure 4 shows the crystal structure of $Nd_2Fe_{14}B$ magnets which is tetragonal in nature [19]. This tetragonal crystal structure of NdFeB are resistance to demagnetization and high uniaxial magnetocrystalline anisotropy attributing to its high magnetic strength.

2.3 Sm Fe N magnets

SmFeN magnets belong to the group of $R_2Fe_{17}N_x$ intermetallic compounds where R represents a rare earth element, and x is typically between 2 and 3. These compounds shows rhombic dodecahedron or hexagonal structure, with nitrogen occupying octahedral interstitial sites. Among all $R_2Fe_{17}N_x$ compounds, $Sm_2Fe_{17}N_x$ is the most promising permanent magnet material due to its uniaxial (c-axis) anisotropy. **Figure 5** presents the $Sm_2Fe_{17}N_3$ (a) rhombohedral and (b) disordered hexagonal structure.

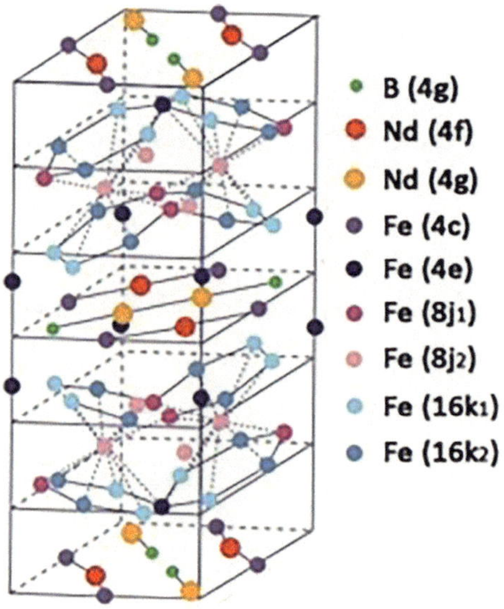

Figure 4.
Crystal structure of $Nd_2Fe_{14}B$ magnets [18].

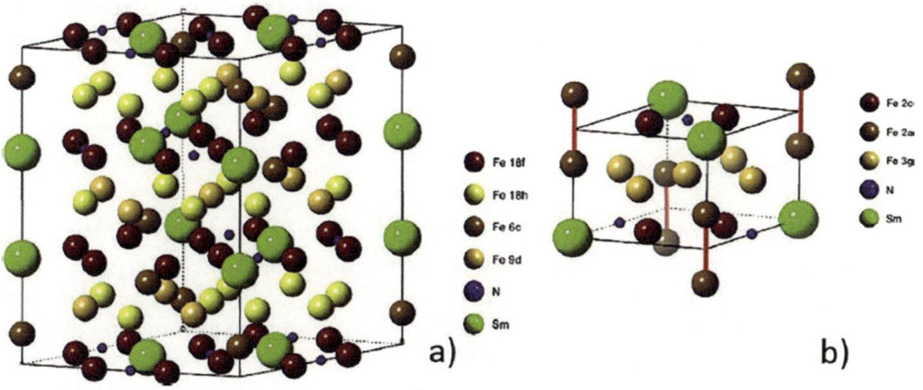

Figure 5.
Structure of samarium iron nitride (a) rhombohedral and (b) hexagonal [20].

At room temperature, Sm_2Fe_{17} compound is a Th_2Zn_{17} type structure [21, 22]. The Th_2Zn_{17} type crystal structure is the basic types of crystal structures of rare-earth permanent magnetic compounds. The spatial stereo diagram of the Th_2Zn_{17} type crystal structure is shown in **Figure 6**, which belongs to the rhombic square crystal system (or tripartite crystal system) with the space group R3m, and a single cell contains three Sm_2Fe_{17} molecules. There are 57 atoms in the unit a cell, 6 Th (or R) atoms occupy the c-site, 9 out of 51 Zn (or Co, Fe, etc.) atoms occupy the d-site, 18 occupy the f-site, 18 occupy the h-site, and 6 occupy the c-site. There are two large interstitial positions in this structure: one is an octahedral interstitial position, located at the 9e crystal site, on the atomic plane containing rare earth atoms; the other is a 3b crystal site located between two rare earth atoms along the

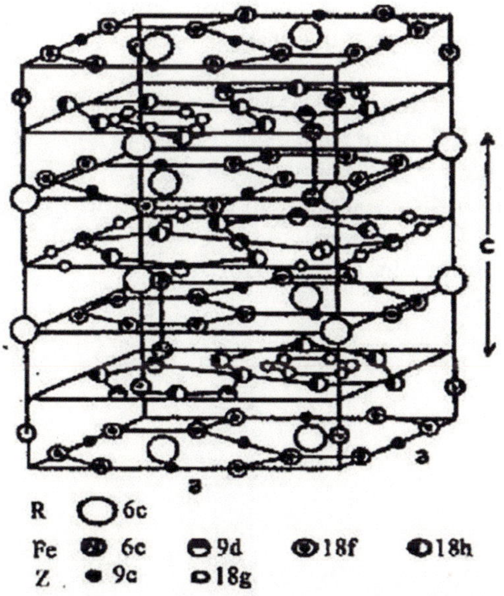

Figure 6.
Single cell structure of 2:17: N compound [23, 24].

c-axis. The $Sm_2Fe_{17}N_x$ [21, 25]-type compounds has the same structure as Sm_2Fe_{17}, with an increase in both a and c, and an increase in the volume of the single cell by about 6%.

3. Properties of rare earth magnets

The magnetic parameters such as high saturation magnetization (M_s), High remanence (M_r), Very high uniaxial magnetocrystalline anisotropy energy (K_1), high coercivity (H_c), High maximum energy product $(BH)_{max}$ and High Curie temperature (T_C) decide the performance of permanent magnets. While the parameters like good temperature stability, mechanical strength, machinability, and low cost determines their applications. The key factor maximum energy product $(BH)_{max}$ treated as the figure of merit for characterization of particular grade permanent magnet.

Figure 7 presents the characterizes B-H loop of the magnets. The shape of the magnet defined the energy product, working point and load line. The mixture of different phases of constituent element in the matrix determines the intrinsic magnetic properties. A good permanent magnet must have a large spontaneous magnetization in zero field (i.e., a high retentivity) and a high coercive force to prevent its being easily demagnetized by an external field. Properties like, density, Curie temperature, Vickers hardness, compress strength, electronic resistivity, bending strength, stretching strength, hot coefficient of expansion and anti-causticity defines the effectiveness of the application of magnets.

Samarium Cobalt based magnets shows exceptional magnetic properties, excellent temperature stability, durability and a high resistance to corrosion or demagnetization [27]. However, 2:17-type SmCo magnets are very sensitive towards the magnetic properties due to the alteration of the magnet composition. For example, if the composition of the Sm_2Co_{17} magnet is changed to Sm $(Co_{bal}Fe_{0.21}Cu_{0.08}Zr_{0.03})_{7.6}$, it affects the coercivity [28]. A step is also observed in the demagnetization curves as shown in **Figure 8**. Further, as observed on this figure, the solution temperature also affects the magnetic order.

The change of composition and materials processing temperature of alloys effect the $(BH)_{max}$ of Sm-Co magnets [29]. Again, $SmCo_7$ phase is always change to 1:5 SmCo phase and 2:17 SmCo due to metastable nature. In order to stable the $SmCo_7$ into $TbCu_7$ structure, another element such as Zr, Hf, Nb, Ti, Ga, etc. are introduced into the matrix to increase its intrinsic magnetic properties [30, 31]. Various element such as Zr are also included for the enhancement of the anisotropy field and the coercivity of $SmCo_7$ powders [17]. The nanostructural modification of the magnet matrix improve the magnetic properties of the magnets.

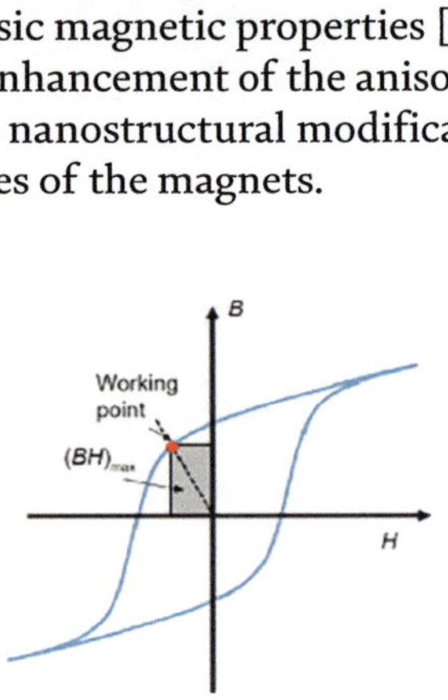

Figure 7.
The B-H loop of a permanent magnet [26].

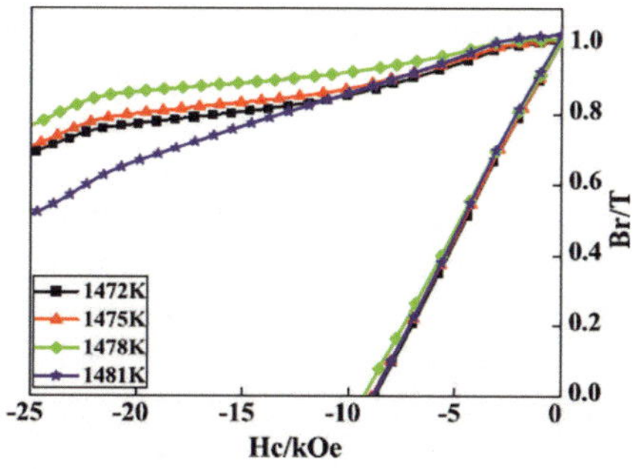

Figure 8.
Demagnetization curves of Sm $(Co_{bal}Fe_{0.21}Cu_{0.08}Zr_{0.03})_{7.6}$ magnets [28].

NdFeB exhibits highest magnetic properties and maximum performance among all permanent magnets. However, temperature has an effect on the performance of the magnets. It performs very well in lower temperature. These magnets offer the highest energy product, higher remanence, much higher coercivity but lower Curie temperature than other types of magnets. The NdFeB magnets may be demagnetised

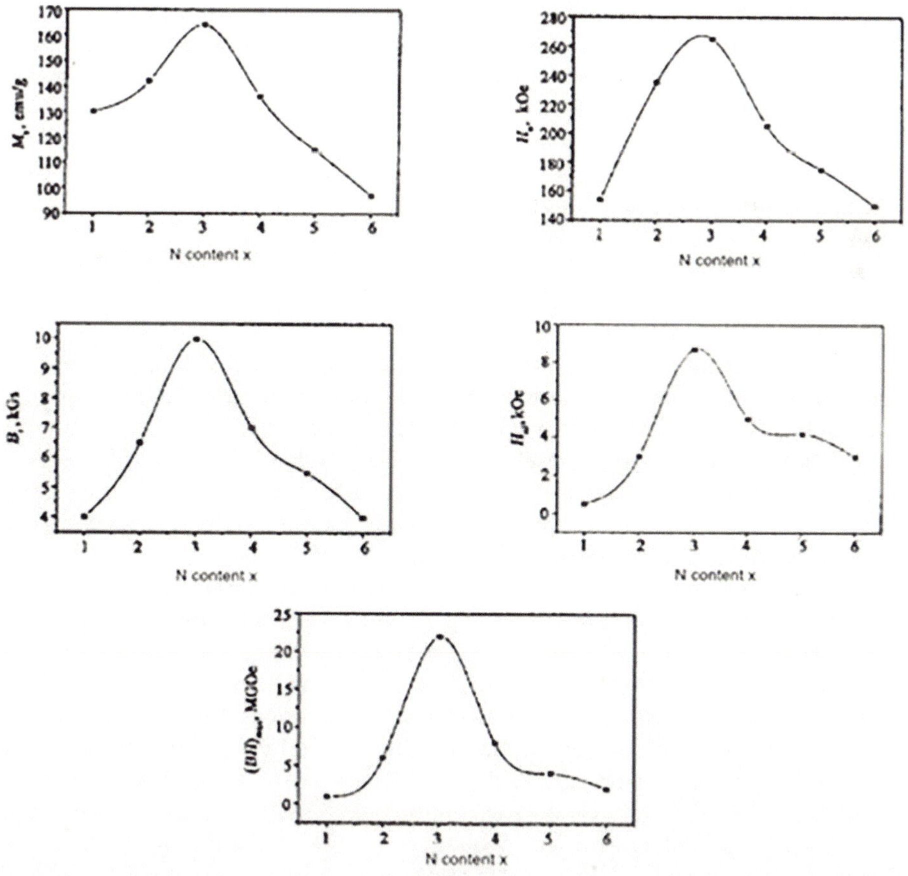

Figure 9.
Magnetic properties of $Sm_2Fe_{17}N_x$ as a function of N content [33, 34].

by radiation [32]. The microstructure with net shape formability and the exchange coupling between $Nd_2Fe_{14}B$ grains manifest better magnetic properties.

Samarium-iron-nitride are considered as high performance and high temperature environment permanent magnets due to its high saturation magnetization value and large magnetocrystalline anisotropy field. Additionally, SmFeN magnets have a theoretical maximum magnetic energy product comparable to that of NdFeB magnets. $Sm_2Fe_{17}N_x$ exhibit excellent magnetic properties including high coercivity field and higher Curie temperature. The N content has a crucial influence on the magnetic properties of $Sm_2Fe_{17}N_x$ compounds [33, 34]. **Figure 9** shows the change in magnetic properties of $Sm_2Fe_{17}N_x$ with the number of N atoms. When the N content x = 6, three N atoms occupy 9e crystal sites, which play an enhanced role in magnetic properties. The other three N atoms occupy half of the 18 g crystalline sites, which weaken the magnetic energy.

Further addition of different metallic elements such as (M = Nb, V, Ta and Co) on Sm_2 (Fe. M) $_{17}$ also increases the Curie temperature of the alloys which is higher than Sm_2Fe_{17} compounds. However, the Curie temperature and spontaneous magnetization intensity of $Sm_2(Fe,M)_{17}N_x$ compound were lower than those of $Sm_2Fe_{17}N_x$. In contrast, for Co, the Curie temperature was increased by nearly 100 K compared to $Sm_2Fe_{17}N_x$ to 845 K and the spontaneous magnetization intensity was 1.41 T. Moreover, the Curie temperature and the magneto-crystal anisotropy field increased with the increase of Co content in a certain range. The addition of Cr and Ga significantly increased the H_{cJ} of $Sm_2(Fe_{1-x}M_x)N_y$ compounds. However, the addition of Zr promoted the amorphization of Sm_Fe alloy and inhibited the grain growth. The addition of Co increased the Curie temperature but decreased the H_{cJ} because $Sm(Fe,Co)_2N_x$ and $Sm(Fe,Co)_3N_x$ were easily formed after the addition of Co interstitial compounds, which decompose below 500°C, resulting in the formation of the soft magnetic phase α(Fe,Co), leading to a decrease in the coercivity [35].

4. Manufacturing process

The manufacturing process is very crucial for production of good permanent magnet to achieve excellent magnetic properties. Even in an opposing magnetic field, a permanent magnet will exhibit magnetic characteristics. The favorable magnetic characteristics such as Curie temperature, coercivity and anisotropy are achieved from the mixtures of the constitute matrix designed for the magnets through structural and thermodynamic modifications. Few known manufacturing processes are used for the fabrication of permanent magnets.

Bonding is the common process used for manufacturing og Ne-Fe-B and Sm-Co magnets. The main processing routes such as calendaring, injection molding, extrusion, and compression bonding are used for manufacturing of most bonded magnets [36]. Other emerging magnet manufacturing processes are extrusion, additive manufacturing, spark plasma sintering, shock compaction, and thermomagnetic processing [37]. Furthermore, rare earth magnets fabricated by rapidly solidified process shows better magnetic properties than conventual processed route. The kinetics and thermodynamics of different processing routes such as melt spinning, atomization and melt extraction also determines the quality and characteristics of the magnets.

Figure 10 shows the standard process steps for manufacturing of the Sm-Co magnets. The steps include alloy preparation, powder production, particle alignment, pressing, sintering, heat treatment, machining, and finally magnetizing. SmCo

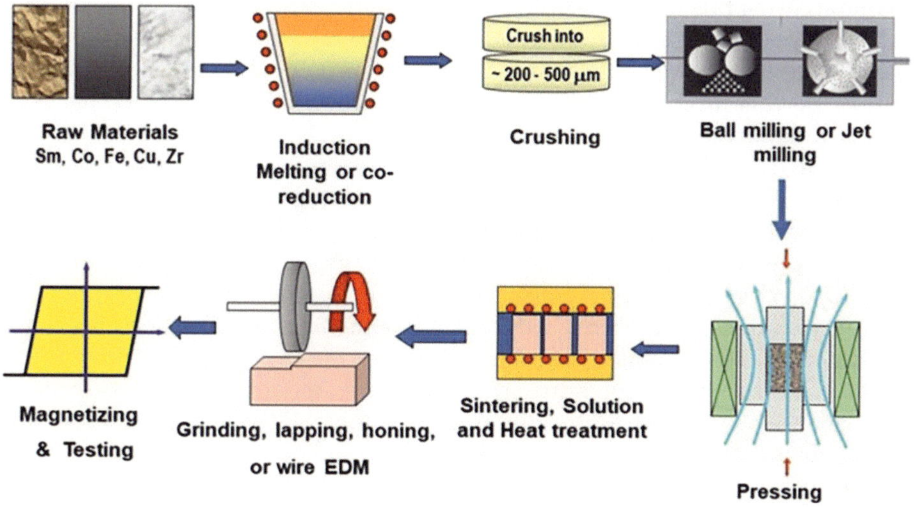

Figure 10.
Manufacturing process steps for Sm-Co based magnets [37, 38].

magnets indicates high temperature stability, which can operate in conditions up to 500°C without suffering any magnetization loss.

The magnets alloy composition and microstructure are the critically important for the processing of NdFeB magnets. A well-defined sintering treatment at a suitable temperature result in high density final product magnets. The basic process step for the manufacture of NeFeB magnet is shown in **Figure 11**.

Mostly sintering, polymer bonding and hot deformation are used as fabrication routes for manufacture of NdFeB-based bulk magnets. In recent days melt quenching nanocrystalline material are used as bonded and hot deformed components and sintering microcrystalline powder.

The SmFeN compounds are thermodynamically metastable and begin to decompose above 600°C, making traditional sintering or hot-pressing methods unsuitable

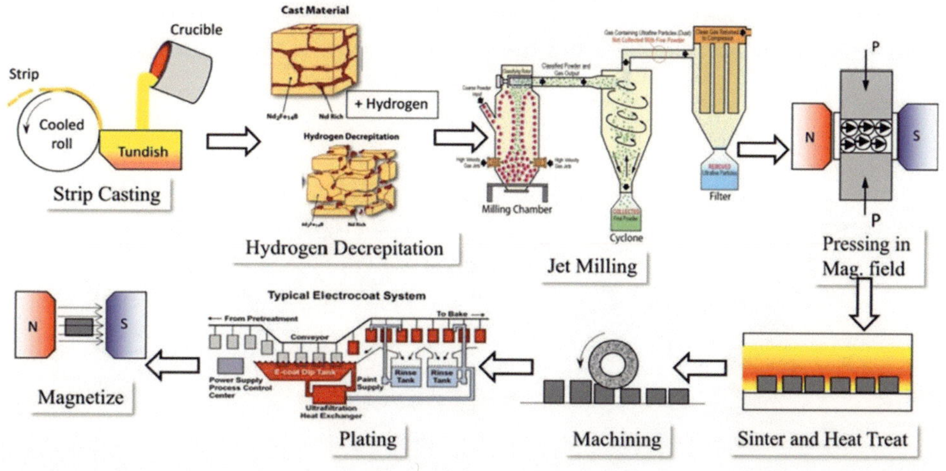

Figure 11.
The manufacturing process steps for the Nd-Fe-B-based magnets [37, 39].

for manufacturing magnets. However, bonded magnets can be produced from SmFeN alloy powders. The alloy powders can be prepared using traditional powder metallurgy methods or by using melt spinning or mechanical alloying techniques to improve powder properties. The preparation of $Sm_2Fe_{17}N_x$ compound magnetic powder is divided into two steps: the first is to prepare the single-phase Sm_2Fe_{17} compound, and the second is to nitride the Sm_2Fe_{17} compound to produce $Sm_2Fe_{17}N_x$. The methods of preparing Sm_2Fe_{17} compound include rapid quenching, reductive diffusion, powder metallurgy and hydrogenation disproportionation. The reduction diffusion method is used for synthesis of Sm-Fe-N powders. The synthesized powders are used in injection molded or compression bonded magnets. Further, the Low temperature milling process are successfully implemented for fabricating the high-performance Sm–Fe–N bonded magnets [40].

In general, a sintered magnet process is used for manufacturing of anisotropic high remanence regular shaped magnet of large volume. However, hot pressed or die-upset magnet are suitable for net-shape formability and powder stability.

4.1 Improvement in properties

The magnetic properties of permanent magnet alloys are always sensitive to the microstructure, and the addition of rare earth elements mainly increases the magnetic crystal anisotropy of the material. The crystal anisotropy is an intrinsic factor for the strong coercivity of rare earth-transition group alloys which enhance the properties of magnets. The Curie temperature, coercivity and magnetization of rare-earth permanent magnets are the primary properties of interest for application. Making intermetallic with suitable doping with metals and non-metals atom for Fe or inserting interstitials atoms that can change lattice site and can significantly increase in Curie temperature. Grain boundary diffusion process is an emerging technology implemented to improve the coercivity of Nd-Fe-B magnets with low rare earth consumption. **Figure 12** shows the coercivity enhancement mechanism by grain boundary optimization. In this process, the grain boundary phase is modified to increase the decoupling of exchange interactions for coercivity enhancement.

It is also reported that the properties of rare earth magnets can be improved through microstructure modification [42]. **Figure 13** shows the mechanism of microstructural modification for the enhancement of coercivity. The coercivity depends on the intrinsic magnetic properties and the microstructure of the magnets. Generally,

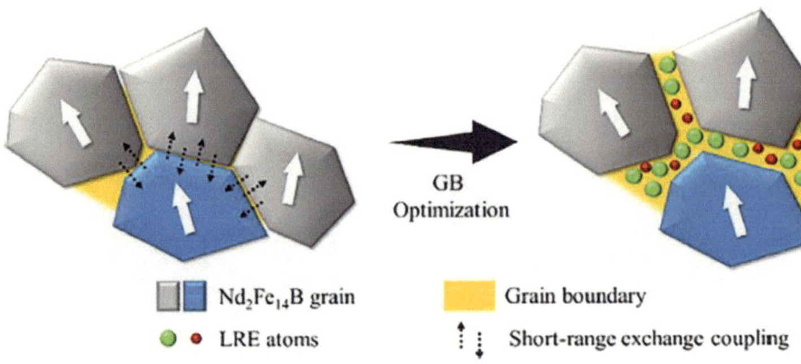

Figure 12.
The schematic diagram of the coercivity enhancement mechanism by GB optimization [41].

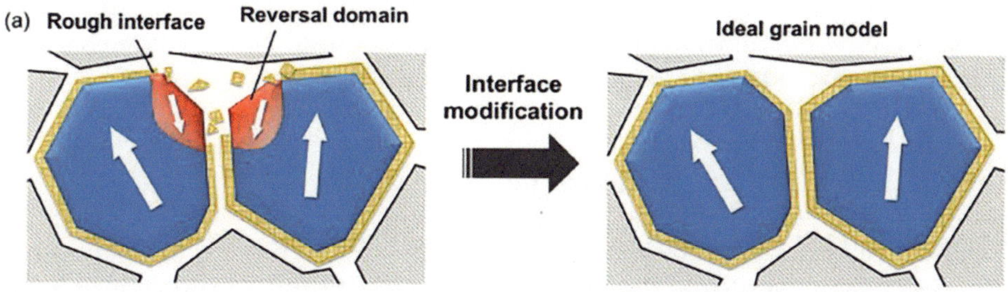

Figure 13.
Coercivity enhancement by microstructural modification [41, 42].

a demagnetizing field is easily forms around the hard magnetic grains in a magnet which reduces the coercivity of the magnet.

Further, the grain surface and the grain boundary are the weak regions in the magnet, where the magnetization reversal starts [43]. Microstructure modification will be performed to retard the magnetic reversal and increase the coercivity. This can be done by the enhancement of the anisotropy field by compositional modification. The defect on the surface of the grain will be reduced through smoothing of grain boundary or optimizing the grain boundary phase distribution. This process will improve the coercivity of the magnet. Grain boundary diffusion process is under test and research for improving the performance rare earth permanent magnets.

Further, introduction of other rare earth elements can modify the intrinsic magnetic properties of magnets. For example, Dy in Nd–Fe–B permanent magnet provide enough coercivity at elevated temperatures by controlling the nanostructures of the materials [44]. The size of the particle alters the magnetic properties of the magnets. The domain structure and chemistry of respective phases of the heterogenous nanostructure can change the coercivity of the magnetic materials. Now a day machine-learning techniques are also used for designing magnetic materials with reduced rare earth components by integrating different physical model [45]. The advancement and improvement in properties of rare-earth will be further done with the addition of nanocrystalline materials.

5. Applications

Permanents magnets have many applications in everyday life from basics to high profile. The specific application of the magnets also depends on the magnetic performance of the materials at a given temperature and cost. Modern permanent magnets are important for academic research, defense applications along with energy technologies. The rare earth magnets have wider range of application in our daily life to modern technology and devices [4].

Bonding applications are common for SmCo magnets. For high temperature application specially in the field of Space, aeronautics and electric vehicle, Sm–Co-based alloys magnets are used [11, 46]. The most powerful Neodymium Rare Earth Magnets has versatile application which includes motors, sensors, mobile phones, computer hard drives and data storage industry [47–50], washing machines, refrigerators. NdFeB magnets markets are the automobile industry and energy sector. SmFeN magnets also have potential applications in electric motors, generators, actuators, magnetic bearings, and other high-performance magnetic devices. Their high coercivity field and high saturation magnetization make them suitable for use in motors and generators that require high torque and power density.

6. Key challenges of rare earth permanent magnets

The performance and properties of magnets varies from magnets to magnet and batch to batch manufacturing. Design and development of suitable processing technique without losing the intrinsic magnetic properties is always a challenge. Further, consumption of critical rare earth elements for high coercivity low-cost magnets, increasing scarcity, price fluctuation and supply instability are the key factor. The cost ratio of rare earth permanent magnet is another challenge. Further the challenges and limitations associated with rare earth magnets are particularly regarding the availability of key rare earth elements.

7. Summary

Rare earth magnets play critical role in many advanced technologies. The rapid growth demand for rare earth magnet is placing increased stress on the availability of key rare earth elements. The improvements to manufacturing processes can form near-net-shape product with reduce materials losses and total machining costs. SmFeN magnets can be an alternative to NdFeB magnets for certain applications. SmFeN can be a new generation magnet after the optimisation of the chemical composition of the elements, the preparation of alloy powder, the coercivity mechanism and the preparation of high-performance magnets with enhanced magnetic properties. A lot of research needs to be done for development of new rare-earth magnetic material magnets for high-temperature applications. Additional experiments or analyses can be conducted to further investigate the development of new rare earth magnetic materials for high temperature applications.

Author details

Dipti Ranjan Sahu
Faculty of Health, Department of Biology, Chemistry and Physics, School of Natural and Applied Sciences, Natural Resources and Applied Sciences, Namibia University of Science and Technology, Windhoek, Namibia

*Address all correspondence to: dsahu@nust.na

References

[1] Kumari A, Sahu SK. A comprehensive review on recycling of critical raw materials from spent neodymium iron boron (NdFeB) magnet. Separation and Purification Technology. 2023;**317**:123527, 1-17

[2] Poudel B, Amiri E, Rastgoufard R, Mirafzal B. Toward less rare-earth permanent magnet in electric machines: A Review. IEEE Transactions on Magnetics. 2021;**57**(9):90019

[3] McCallum RW, Lewis LH, et al. Practical aspects of modern and future permanent magnets. Annual Review of Materials Research. 2014;**44**:451-477

[4] Brown D, Ma B-M, Chen Z. Developments in the processing and properties of NdFeB-type permanent magnets. Journal of Magnetism and Magnetic Materials. 2002;**248**(3):432-440

[5] Liu H, Zhang Y, Luan Y, Yu H, Li D. Research progress in preparation and purification of rare earth metals metals. 2020;**10**(10):1376, 1-13

[6] Filho WL, Kottor R, et al. Understanding rare earth elements as critical raw materials. Sustainability. 2023;**15**(9):1919, 1-18

[7] Binnemans K, Tom P, Blanpain B, Gerven TV, Yang Y, Walton A, et al. Recycling of rare earths: A critical review. Journal of Cleaner Production. 2013;**51**:1-22

[8] Zhou X, Huang A, Cui B, Sutherland JW. Techno-economic assessment of a novel SmCo permanent magnet manufacturing method. Procedia CIRP. 2021;**98**:127-132

[9] Coey JMD. Perspective and prospects for rare earth permanent magnets. Engineering. 2020;**6**(2):119-131

[10] Horikawa T, Yamazaki M, Mstsuura M, Sugimoto S. Recent progress in the development of high-performance bonded magnets using rare earth–Fe compounds. Scientific Technolog. Advanced Materials. 2021;**22**(1):729-747

[11] Sayama J, Asahi T, Mizutani K, Osaka T. Newly developed $SmCo_5$ thin film with perpendicular magnetic anisotropy. Journal of Physics D: Applied Physics. 2004;**37**:L1-L4

[12] Song XY, Zhang ZX, Lu ND, et al. Crystal structures and magnetic performance of nanocrystalline Sm-Co compounds. Materials Science. 2012;**6**:207-215

[13] Zhang ZX, Song XY, Xu WW, et al. Crystal structure and magnetic performance of single-phase nanocrystalline $SmCo_7$ alloy. Scripta Materials. 2010;**62**:594-597

[14] Landa A, Soderlind P, Parker D, et al. Thermodynamics of SmCo5 compound doped with Fe and Ni: An ab initio study. Journal of Alloy Compound. 2018;**765**:659-663

[15] Guo K, Lu H, Xu GJ, et al. Hierarchically nanostructured $ZnCo_2O_4$ particles in 3D graphene networks for high-rate and long-life lithium ion batteries. Materials Today Chemistry. 2022;**25**:10098, 1-27

[16] Sharma S, Zintler A, et al. Epitaxy induced highly ordered sm_2co_{17}–$smco_5$ nanoscale thin-film magnets. ACS Applied Materials & Interfaces. 2021;**13**(27):32415-32423

[17] Das B, Choudhary R, Skomski R, et al. Anisotropy and orbital moment in Sm-Co permanent magnets. Physical Review B. 2019;**100**:024419

[18] Givord D, Li HS, Moreau JM. Magnetic properties and crystal structure of Nd2Fe14B. Solid State Communication. 1984;**50**(6):497-499

[19] Lewis LH, Villacorta FJ. Perspectives on permanent magnetic materials for energy conversion and power generation. Metallurgical Materials Transaction A. 2013;**44**:2020

[20] Coey JM, Iriyama T. Bonded Sm-Fe-N Magnets: Modern Permanent Magnets. Woodhead Publishing Series in Electronic and Optical Materials. United Kingdom; 2022. pp. 305-342

[21] Yan W, Luo Y, Wu G, et al. Controlled magnetic properties by tuning TbCu7/Th2Zn17 phase in isotropic Sm-Fe-Nb-N compounds. Journal of Alloys and Compounds. 2018;**74**:661-665

[22] Lotfi B. Structure and Magnetic Properties of Intermetallic Rare-Earth-Transition-Metal Compounds: A Review. Materials. 2022;**15**:201-223

[23] Ainai Y, Shiozawa T, Tatetsu Y, Gohda Y. First-principles study on surface stability and interface magnetic properties of $SmFe_{12}$. Applied Physics Express. 2020;**13**:045502

[24] Li W, Gau J, Zuo S, Wang Y, Huang G, Zheng L. Rizinia.com, Research Progress of SmFeN rare earth permanent magnetic materials. 2022

[25] Yamg YC, Zhang XD, Ge SL, Pan Q, Kong LS, Li H. Magnetic and crystallographic properties of novel Fe-rich rare-earth nitrides of the type RTiFe11N1- δ. Journal of Applied Physics. 1991;**70**:6001-6005

[26] Coey JMD. Perspective and prospects for rare earth permanent magnets. Engineering. 2020;**6**:119-121

[27] Guo K, Lu H, Xu GJ, et al. Recent progress in nanocrystalline Sm–Co based magnets. Materials Today Chemistry. 2022;**25**:100983

[28] Zhang TL, Song Q, Wang H, et al. Effects of solution temperature and Cu content on the properties and microstructure of 2: 17-type SmCo magnets. Journal of Alloys and Compounds. 2028;**735**:197-1976

[29] Dormidontov AG, Kolchugina NB, Dormidontov NA, Milov YV. Structure of Alloys for (Sm,Zr)(Co,Cu,Fe)Z Permanent Magnets: First Level of Heterogeneity. Materials. 2002;**13**:3893, 1-18

[30] Landa A, Soderlind P, Parker D, et al. Thermodynamics of $SmCo_5$ compound doped with Fe and Ni: An ab initio study. Journal of Alloys and Compounds. 2018;**765**:659-661

[31] Yin SQ, Wang H, Zhao HB, et al. The effects of Cu doping on crystalline structure and magnetic properties of $SmCo_{5-x}Cu_x$ thin films grown on Ru (0002). Journal of Applied Physics. 2013;**114**:213908

[32] Samin A. A review of radiation-induced demagnetization of permanent magnets. Journal of Nuclear Materials. 2018;**503**:42-55

[33] Iriyama T, Kobayashi N, Imaoka, et al. The origin of the enhancement of magnetic properties in $Sm_2Fe_{17}N_x$ ($0< x\lesssim 3$). Journal of Magnetics Materials. 2002;**247**:42-54

[34] Wei YN, Sun K, Feng YB. Structural and intrinsic magnetic properties of $Sm_2Fe_{17}N_y$ (y= 2-8). Journal of Alloys and Compounds. 1993;**194**:9-12

[35] Zhang LLG, Shen BG, Zhang SY, Zhang HW. Magnetocrystalline anisotropy of $Er_2(Fe_{1-x}Cox)_{15}Ga_2$

compounds. Acta Physics Sinica. 1998;**47**(5):817-823

[36] Cui J, Ormero J, et al. Manufacturing processes for permanent magnets: Part II—bonding and emerging methods. JOM. 2022;**74**(6):2492-2506

[37] Cui J, Ormero J, et al. Manufacturing processes for permanent magnets: Part I—sintering and casting. JOM. 2022;**74**(4):1279-1295

[38] Liu J. Sm-Co Permanent Magnets and Their Applications. Dayton Magnetics Seminar, Dayton, OH; 2008

[39] Ormerod J. Rare Earth Magnets: Yesterday, Today and Tomorrow. Orlando, USA: Magnetic Applications, Inc; 2019

[40] Saito J. Magnetic properties of anisotropic Sm–Fe–N bulk magnets produced by spark plasma sintering method. Journal of Magnetism and Magnetic Materials. 2008;**320**:1893-1897

[41] Zhongwu L, He J, Ramanujan RV. Significant progress of grain boundary diffusion process for cost-effective rare earth permanent magnets: A review. Materials & Design. 2021;**209**:110004

[42] Soderznik M, Sepehri-Amin H, Sasaki TT, et al. Magnetization reversal of exchange-coupled and exchange-decoupled Nd-Fe-B magnets observed by magneto-optical Kerr effect microscopy. Acta Materialia. 2017;**135**:68-76

[43] Xu XD, Sepehri-Amin H, Sasaki TT, et al. Comparison of coercivity and squareness in hot-deformed and sintered magnets produced from a Nd-Fe-B-Cu-Ga alloy. Scripta Materialia. 2019;**160**:9-14

[44] Hirosawa S, Matsuura Y, Yamamoto H, et al. Magnetization and magnetic anisotropy of R2Fe14B measured on single crystals. Journal of Applied Physics. 1986;**59**(3):873-879

[45] Kovacs A, Fischbacher J, Oezelt H, et al. Physics-informed machine learning combining experiment and simulation for the design of neodymium-iron-boron permanent magnets with reduced critical-elements content. Frontiers in Materials. 2023;**9**(1094055):1-19

[46] Jiang X. Structural, magnetic and microstructural studies of composition-modified Sm-Co ribbons. Mechanical (and Materials) Engineering – Dissertations, Theses, and Student Research. 2014:68

[47] Wang H, Lamichhane TN, Paranthaman MP. Review of additive manufacturing of permanent magnets for electrical machines: A prospective on wind turbine. Materials Today Physics. 2022;**24**:100675

[48] Gutfleisch O, Willard MA, Brück E, et al. Magnetic materials and devices for the 21st century: Stronger, lighter, and more energy efficient. Advanced Materials. 2011;**23**(7):821-842

[49] Yoshida Y, Yoshikawa N. Modern Permanent Magnets. United Kingdom: Woodhead Publishing Series in Electronic and Optical Materials; 2022. pp. 251-304

[50] Trench A, Sykes JP. Rare earth permanent magnets and their place in the future economy. Engineering. 2020;**6**(2):115-118